Springer Theses

Recognizing Outstanding Ph.D. Research

Aims and Scope

The series "Springer Theses" brings together a selection of the very best Ph.D. theses from around the world and across the physical sciences. Nominated and endorsed by two recognized specialists, each published volume has been selected for its scientific excellence and the high impact of its contents for the pertinent field of research. For greater accessibility to non-specialists, the published versions include an extended introduction, as well as a foreword by the student's supervisor explaining the special relevance of the work for the field. As a whole, the series will provide a valuable resource both for newcomers to the research fields described, and for other scientists seeking detailed background information on special questions. Finally, it provides an accredited documentation of the valuable contributions made by today's younger generation of scientists.

Theses are accepted into the series by invited nomination only and must fulfill all of the following criteria

- They must be written in good English.
- The topic should fall within the confines of Chemistry, Physics, Earth Sciences, Engineering and related interdisciplinary fields such as Materials, Nanoscience, Chemical Engineering, Complex Systems and Biophysics.
- The work reported in the thesis must represent a significant scientific advance.
- If the thesis includes previously published material, permission to reproduce this must be gained from the respective copyright holder.
- They must have been examined and passed during the 12 months prior to nomination.
- Each thesis should include a foreword by the supervisor outlining the significance of its content.
- The theses should have a clearly defined structure including an introduction accessible to scientists not expert in that particular field.

More information about this series at http://www.springer.com/series/8790

Valerio Rossetti

Search for Exotic Mono-jet Events

in Proton–Proton Collisions at $\sqrt{s} = 7$ TeV and 8 TeV with the ATLAS Detector

Doctoral Thesis accepted by
the University of Barcelona, Spain

Springer

Author
Dr. Valerio Rossetti
Stockholm University
Stockholm
Sweden

Supervisor
Prof. Mario Martinez Perez
ICREA IFAE/UAB Campus Bellaterra
Barcelona
Spain

ISSN 2190-5053 ISSN 2190-5061 (electronic)
Springer Theses
ISBN 978-3-319-22224-0 ISBN 978-3-319-22225-7 (eBook)
DOI 10.1007/978-3-319-22225-7

Library of Congress Control Number: 2015947107

Springer Cham Heidelberg New York Dordrecht London

Printed on acid-free paper

Springer International Publishing AG Switzerland is part of Springer Science+Business Media
(www.springer.com)

Publications Related to this Thesis

The results presented in this thesis led to the following publications by the ATLAS collaboration:

- *Search for new phenomena with the monojet and missing transverse momentum signature using the ATLAS detector in* $\sqrt{s} = 7$ *TeV proton–proton collisions*, Phys. Lett. B705 (2011) 294–312, arXiv:1106.5327 [hep-ex] (based on 33 pb^{-1} of data collected in 2010)
- *Search for New Phenomena in Monojet plus Missing Transverse Momentum Final States using 1* fb^{-1} *of pp Collisions at* $\sqrt{s} = 7$ *TeV with the ATLAS Detector*, ATLAS-CONF-2011-096, http://cds.cern.ch/record/1369187 (based on 1 fb^{-1} of data collected in 2011)
- *Search for dark matter candidates and large extra dimensions in events with a jet and missing transverse momentum with the ATLAS detector*, JHEP 04 (2013) 075, arXiv:1210.4491 [hep-ex] (based on 4.7 fb^{-1} of data collected in 2011)
- *Search for New Phenomena in Monojet plus Missing Transverse Momentum Final States using 10* fb^{-1} *of pp Collisions at* $\sqrt{s} = 8$ *TeV with the ATLAS detector at the LHC*, ATLAS-CONF-2012-147, http://cds.cern.ch/record/1493486 (based on 10.5 fb^{-1} of data collected in 2012).

Appendix F includes a discussion on a complementary mono-photon analysis led by IFAE and published in Phys. Lett. Rev. 110 (2013) 011802. Finally, the appendices also include contributions to the Tile calorimeter commissioning, leading to the publication in Eur. Phys. J. C 70 (2010) 1193–1236.

At the time of the publication of this manuscript, the ATLAS collaboration published other two analyses on searches with monojet final states, and using the full 8 TeV dataset. One was published in Phys. Rev. D. 90, 052008 (2014), and another was submitted to EPJC. The main results from these publications are summarized in Appendix G, at the end of this book.

A nonna Maria,
che ha sempre incoraggiato figli e nipoti
a proseguire nei loro studi.
Quegli studi che lei non ha potuto,
ma ha sempre voluto fare.

To grandma Maria,
who always encouraged her children
and grandchildren to carry on
with their studies.
The studies that she could not,
but wanted to pursue.

Supervisor's Foreword

This thesis is the result of more than four years in the exploration of the secrets of Nature using high-energy proton–proton collisions at the Large Hadron Collider (LHC) at CERN, collected by the ATLAS experiment in the years 2010–2012.

Since the late 1970s, the search for new phenomena using mono-jet events has attracted the attention of the scientific community. Such a simple final state, with one jet of hadrons and nothing else, could turn crucial to reveal the presence of new phenomena. Already in the early 1980s at the UA1 experiment at CERN (Geneva, Switzerland) Prof. C. Rubbia (co-winner of the 1984 Nobel Prize for the discovery of the W as Z bosons) believed for a moment that his mono-jet data showed the first signals for the production of dark matter or supersymmetry. However, this was not confirmed by other experiments, and it was finally understood in terms of the irreducible Z (→neutrinos)+jets background, a difficult-to-model background process.

No doubt the nature of dark matter is one of the big open questions in particle physics nowadays. The observations of the orbital velocities of stars in galaxies and galaxies in clusters show a discrepancy between the total mass, as inferred from their gravitational effects, and that calculated from the observed distribution of light. Such a discrepancy is known since the 1930s and cannot be accommodated with nonluminous ordinary matter. In order to account for this "missing" mass, physicists postulate the existence of dark matter, a new kind of matter that would not emit or absorb light. Other observations have added evidence for the presence of dark matter in the Universe. This includes the gravitational lensing of objects by galaxy clusters, the study of cluster–cluster collisions, and the pattern of anisotropies in the cosmic microwave background. In fact, our galaxy would be surrounded by a large halo of dark matter, and dark matter would make up around 85 % of the Universes mass.

The favorite candidates for dark matter are WIMPs, or weakly interacting massive particles. The mono-jet analysis has been always considered the golden channel for the search for WIMPs at colliders. Mono-jet final states also provide unique access to the search for supersymmetry in very mass compressed scenarios that involve the presence of neutralinos in the final state, considered as the preferred dark matter candidates in many supersymmetric models.

Moreover, in 1998 new ideas by N. Arkani-Hamed, S. Dimopoulos, and G. Dvali to explain the weakness of gravity led to the revolutionary concept of Large Extra Spatial Dimensions (LED). In their model, the graviton (the particle acting as mediator of the gravitational interaction) could be produced in association with a jet of hadrons. The graviton propagates in the extra dimensions, thus explaining the apparent weakness of gravity at subatomic scales, and only one jet of hadrons would be observed in our four-dimensional world. This anticipated, once again, a large excess of events in mono-jet signals at colliders. The CDF and D0 experiments at the Fermilabs proton–(anti)proton collider (Tevatron, near Chicago, USA) searched for such a signature before the LHC era.

In 2009, the LHC initiated operations and the searches for new phenomena in mono-jet final states came back as a flagship in the physics programs of the LHC experiments. A protagonist of the mono-jet adventure within the ATLAS Collaboration, a young and brave scientist, writes this book. It collects all the results obtained by the ATLAS experiment to date on the subject. It presents in detail the necessary theoretical foundations and all the experimental aspects behind an important measurement leading to several publications in journals. Up-to-date results on searches for WIMPS, LEDs, and supersymmetric models, leading to mono-jet final states, are compiled in a coherent way. It was commonly believed that jet-related measurements at colliders could not achieve experimental precisions better than 10–20 %. This books will show you how a carefully designed event selection strategy, together with the use of data-driven techniques in constraining the irreducible backgrounds, translates into extremely precise background predictions. Ultimately, this results in an unforeseen sensitivity in the search for new phenomena.

At the time of writing this text, the LHC, after two years of shutdown devoted to improvement in both the collider and the experiments, is about to resume proton–proton collisions with an increased centre-of-mass energy of 13 TeV. A new era on the searches for new fundamental physics is about to open and the mono-jet searches will continue to play a central role. In such a scenario, this book should be regarded as a reference for Ph.D. students, young scientists, and interested readers in general.

Barcelona, Spain
July 2015

Prof. Mario Martinez
ICREA Research Professor
Institut de Físca d'Altes Energies (IFAE)
Head of IFAE Experimental Division
and PI of the ATLAS-IFAE
Research Group

Preface

This thesis presents a search for new phenomena in proton–proton collisions at $\sqrt{s} = 7$ TeV and 8 TeV recorded with the ATLAS experiment at the LHC collider. The final state under investigation is defined by the presence of one energetic jet and large missing transverse momentum. Events with such a final state constitute a clean and distinctive signature at colliders, and are studied to test various scenarios for physics beyond the Standard Model (SM), such as Large Extra Dimensions, Supersymmetry, and the production of Weakly Interacting Massive Particles, among others. In order to achieve the highest sensitivity to new physics signals, data-driven techniques are employed to estimate the background contributions.

This thesis is organized as follows. Chapter 1 gives an introduction to the SM theory, the QCD phenomenology at hadron colliders, and to several scenarios for physics beyond the SM. The LHC collider and the ATLAS experiment are described in Chap. 2. Chapter 3 details the search for new physics in mono-jet events based on 4.7 fb^{-1} of 7 TeV collisions collected in 2011. Chapter 4 describes the results based on 10.5 fb^{-1} of 8 TeV collisions recorded in 2012. Finally, Chap. 5 is devoted to conclusions.

The document is complemented with several appendices. Performance studies on the Tile hadronic calorimeter, and the luminosity measurement with its data, are described in appendices A and B, respectively. Previous results on mono-jets based on 33 pb^{-1} and 1 fb^{-1} of 7 TeV collisions are presented in Appendices C and D. Appendix E includes additional information on the analysis at 7 TeV. The main results on a mono-photon search are collected in Appendix F, and are regarded as a complementary study to the mono-jet analysis. Finally, Appendix G at the end of the book documents the analyses published by ATLAS using the full 2012 dataset (20 fb^{-1}).

Acknowledgments

First, I would like to thank my thesis director, Mario Martinez Perez, for his scrupulous and professional supervision, and for the attention he dedicated to my formation as a physicist. Thanks to Jalal Abdallah, with whom I worked closely for the last three years, and who taught me things on a broad spectrum of subjects, such as physics, gastronomy, and enology. Thanks to Luca Fiorini and Ilya Korolkov, who supervised the activities I have been carrying on the Tile Calorimeter. Thanks to the "Institut de Fisica d'Altes Energies" (IFAE) for the support and encouragements, and for the opportunity I was given to work at CERN. It is worth to mention that this thesis has been carried out also thanks to an FPU fellowship (AP2008-04396) of the "Ministerio de Educación y Ciencia."

Thanks to my officemates, at CERN and at IFAE, for the lovely time we spent together, and for the useful interactions we had. I would like to thank (in order of appearance): Matteo Volpi, Estel Perez, Francesc Vives, Monica D'Onofrio, Evelin Meoni, Roger Caminal, Garoe Gonzalez, Vincent Giangiobbe, Paolo Francavilla, Silvia Fracchia, Antonella Succurro, and Jordi Nadal.

Here I would also like to mention that I am Italian, and I never voted for Berlusconi. In fact, I hate that guy, and I wish him all the worst! His system of power, based on corruption and on the support of criminal organizations, is the main cause for nearly twenty years of cultural and economical regression in my country.

Then, I would like to thank all the friends with whom I spent time during my Ph.D., starting from the components of the band "The Suspenders" (Alexandros, Marco, Matias, Samir, Georgios, and Panos) and the flatmates of the apartment in rue de la Servette 55 (Maria-Elissavet, Matias, Sara, Matteo, Yisel, Samir, Silvia and Cian). It is also worth to mention the incredible achievements of the football team "Los Mariachis," led by Leonardo di Matteo and Ali Abdallah (Jalal's brother), and in which I was honored to have played.

I want to thank Mario Dipoppa, my inseparable companion of adventures, with whom I spent amazing summer vacations. Thanks to Francesca, Maria Elisa, Noemi, Simone, and Stefano for the awesome moments we lived together in these years.

I would like also to spend a few words on the “Pachinko,” a legendary bar in the neighborhood of Grottes in Geneva, that exceptionally opens only on wednesday night. There, I sensibly improved my French language skills and my knowledge of the human behavior in a highly dense environment.

Contents

Chapter 1
Introduction to Theory and Phenomenology

This chapter describes the theoretical and phenomenological aspects relevant for this thesis. In Sect. 1.1, a brief introduction to the SM theory and to Quantum Cromo-Dynamics (QCD) is given. Sections 1.2 and 1.3 deal with QCD phenomenology and Monte Carlo (MC) simulation at hadron colliders. The final part of the chapter is devoted to the physics beyond the SM. Three scenarios for new physics are presented, and their theoretical motivation and their final state signatures at hadron colliders are described.

1.1 Introduction to the Standard Model

The Standard Model (SM) of particle physics [1–3] is a renormalizable quantum field theory that describes the properties of all known fundamental particles (listed in Table 1.1) and their interactions through the electromagnetic, weak and strong forces.

The SM lagrangian is invariant under the gauge transformations of the symmetry group $SU(3)_C \times SU(2)_L \times U(1)_Y$. This invariance results in the existence of elementary interactions mediated by the following vector bosons: the photon (for the electromagnetic interaction), the Z^0 and the $W^{\pm}$ (for the weak interaction), and eight types of the gluons (for the strong interaction). The SM classifies the twelve elementary fermions into two categories, the leptons and the quarks, each of them divided into three generations. Both quarks and leptons interact weakly, while only quarks interact strongly. Each particle has an antiparticle which differs by the sign of the quantum numbers.

In the SM, the only elementary particle with spin zero is the Higgs boson [8–10]. The interaction of the Higgs field with quarks, charged leptons, and with the Z and W bosons, results in lagrangian mass terms for all massive particles. The ATLAS and CMS experiments at CERN discovered a new particle; first measurements of its

V. Rossetti, *Search for Exotic Mono-jet Events*, Springer Theses,
DOI 10.1007/978-3-319-22225-7_1

Table 1.1 List of known fundamental particles with corresponding electrical charge, spin, and mass

List of known fundamental particles				
Family	Name	Charge (e)	Spin [$\hbar$]	Mass (MeV)
Quarks	Up (u)	+2/3	1/2	$2.3^{+0.7}_{-0.5}$
	Down (d)	−1/3	1/2	$4.8^{+0.7}_{-0.3}$
	Strange (s)	+2/3	1/2	95 ± 5
	Charm (c)	−1/3	1/2	1275 ± 25
	Bottom (b)	+2/3	1/2	4180 ± 30
	Top (t)	−1/3	1/2	$(173.5 \pm 1.0) \times 10^3$
Leptons	Electron (e)	−1	1/2	$0.510998928 \pm 0.000000011$
	Electron neutrino (ν_e)	0	1/2	$<2 \times 10^{-6}$
	Muon (μ)	−1	1/2	$105.6583715 \pm 0.0000035$
	Muon neutrino (ν_μ)	0	1/2	<0.19
	Tau (τ)	−1	1/2	1776.82 ± 0.16
	Tau neutrino (ν_τ)	0	1/2	<18.2
Gauge bosons	Photon (γ)	0	1	0
	Z	0	1	$(91.1876 \pm 0.0021) \times 10^3$
	$W^\pm$	±1	1	$(80.385 \pm 0.015) \times 10^3$
	Gluon (g)	0	1	0
Higgs boson (H)		0	0	$(125.5^{+0.5}_{-0.6}) \times 10^3$

The lepton, quarks and gauge bosons masses are taken from the fitted average given by the Particle Data Group [4]. In the last row the new particle discovered by ATLAS [5] and CMS [6] is assumed to be the SM Higgs boson, with the mass measured by ATLAS [7]

properties (spin, mass, parity and charge conjugation, production cross section, and decay branching fractions[1]) favor the interpretation in terms of the SM Higgs boson.

Quantum Chromodynamics (QCD) [11] is the renormalizable gauge field theory that describes the strong interactions between quarks. The QCD lagrangian reads:

$$\mathcal{L}_{QCD} = \sum_{flavor} \bar{q}\,(i\gamma^\mu D_\mu - m_q)\,q - \frac{1}{4} F_A^{\alpha\beta} F^A_{\alpha\beta}, \tag{1.1}$$

where m_q are the masses of the six types of quarks, and q are the quark fields. The covariant derivative D_μ is defined as:

$$D_\mu = \partial_\mu + i g_s G^\alpha_\mu t^\alpha \tag{1.2}$$

[1] Assuming no branching fraction for Higgs decays to exotic particles.

The field tensor $F^A_{\alpha\beta}$ is derived from the gluon field G^A_α:

$$F^A_{\alpha\beta} = \partial_\alpha G^A_\beta - \partial_\beta G^A_\alpha - gf^{ABC} G^B_\alpha G^C_\beta, \tag{1.3}$$

where f^{ABC} are the structure constants of the SU(3) group, and A, B, and C are indices of the eight degrees of freedom of the gluon field. The third term of the tensor is responsible for the characteristic non-abelian nature of the QCD, and it describes the gluon self-interaction. This leads to a particular phenomenology driven by the strength of the strong coupling, $\alpha_s = g/4\pi$, that is large at low energies (large distances), and small at high energies (short distances), as it is shown in Fig. 1.1.

As a consequence, when two quarks separate the field energy increase, and this leads to the quark confinement in hadrons. If two quarks separate far enough, the field energy increases so much that new quarks are created from the vacuum forming colorless hadrons. For this reason quarks cannot be observed as free particles. At small distances, instead, the strength of the QCD coupling is so low, that quarks and gluons can be described essentially as free particles, with $\alpha_s \ll 1$. This phenomenon is named asymptotic freedom, and allows to use the perturbative approach in this regime.

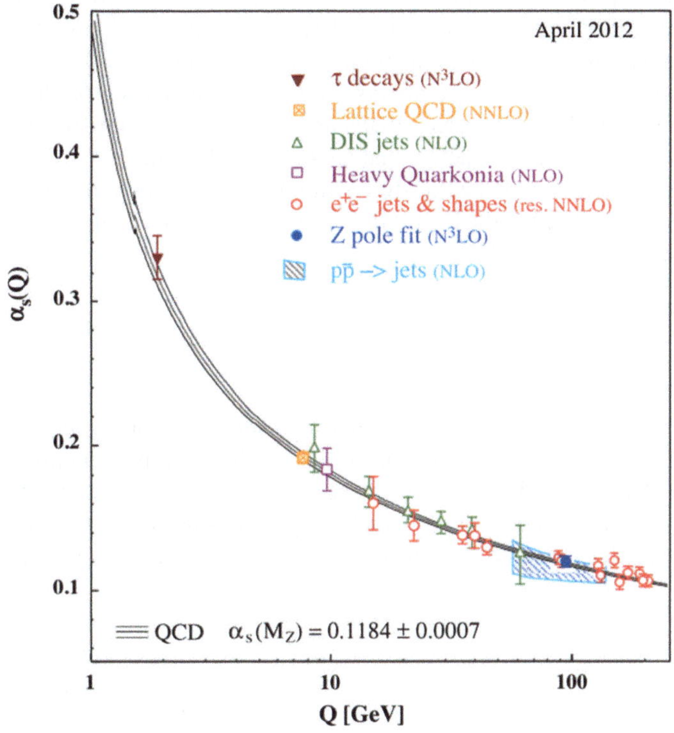

Fig. 1.1 Running of the strong coupling α_s with the energy scale Q, proven from different measurements. The plot is taken from Ref. [12]

1.2 QCD Phenomenology at Hadron Colliders

The QCD theory and its phenomenological implications have a crucial role in hadron collider physics, since it involves both the initial and final states in the collisions. In this section, few relevant aspects of the QCD phenomenology are introduced.

1.2.1 The Proton Structure and the Parton Density Functions

The proton is described as a bound state of three quarks, called "valence quarks", each carrying a fraction of the proton momentum x. Being $f_i(x)dx$ the probability of finding a parton of type i carrying a fraction of the proton's momentum between x and $x + dx$, the following equation has to be satisfied:

$$\int_0^1 x \sum_i f_i(x)dx = 1, \tag{1.4}$$

where the functions $f_i(x)$ are called Parton Density Functions (PDFs). The functional form of the PDFs cannot be predicted in QCD, and is measured in hadron-hadron, and deep inelastic e, μ, and ν scattering. Experimentally it is found that the quarks carry about half of the proton momentum, while the rest is carried by virtual gluons, that are exchanged between the quarks. Gluons, on their side, produce virtual $q\bar{q}$ pairs that are called "sea quarks". PDFs are parametrized with specific functional forms with a set of free parameters that are tuned to the data.

The PDF evolution with the transferred momentum Q^2 can be predicted by perturbative QCD (pQCD) using the DGLAP (Dokshitzer-Gribov-Lipatov-Altarelli-Parisi) equation, which at LO reads:

$$\frac{df_i(x, Q^2)}{dQ^2} = \sum_j \frac{\alpha_s(Q^2)}{2\pi} \int_x^1 \frac{dz}{z} P_{j\to i}(z) f_j(x/z, Q^2). \tag{1.5}$$

$P_{j\to i}(z)$ are called splitting functions and represent the probability that a parton j radiates a parton i, carrying a fraction z of its longitudinal momentum.

Figure 1.2 shows the PDFs of the valence quarks of the proton, the gluon, and the sea quarks for two values of Q^2, at which the proton is probed. As can be seen in the figure, the valence quarks dominate at large x, while the gluon dominates at low x.

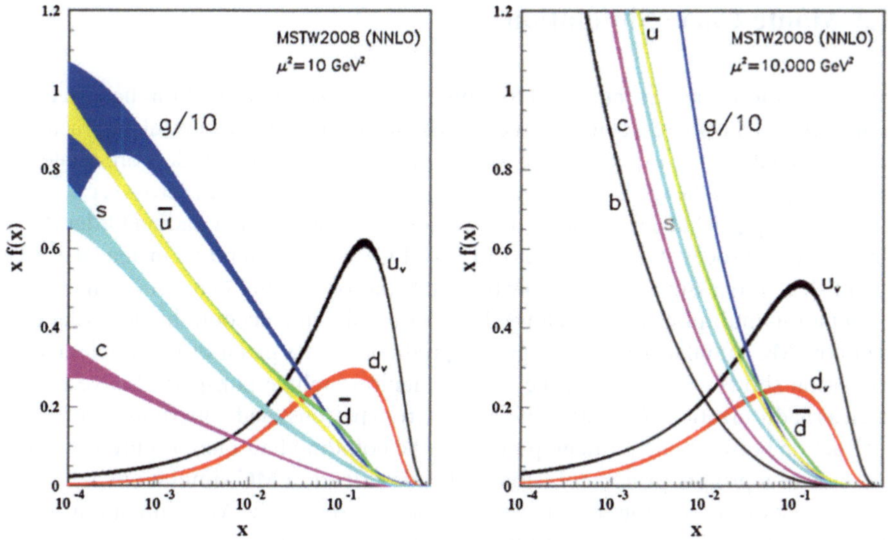

Fig. 1.2 Proton PDF functions at transfer momentum $Q^2 = 10$ GeV, on the *left*-hand side, and $Q^2 = 10{,}000$ GeV, on the *right*-hand side. In the figure $\mu^2 \equiv Q^2$

1.2.2 Factorization Theorem

The QCD factorization theorem is a crucial concept of QCD. It states that cross sections of inclusive processes can be separated into a hard partonic cross section (short-distance) component and a long-distance component:

$$\sigma(P_1, P_2) = \sum_{i,j} \int dx_1 dx_2 f_i(x_1, \mu_F^2) f_j(x_2, \mu_F^2) \times \sigma_{ij}(x_1, x_2, \alpha_s(\mu_F^2, \mu_R^2), Q^2/\mu_F^2), \quad (1.6)$$

where P_1, P_2 are the momenta of the interacting hadrons, the sum runs over all parton types, and σ_{ij} is the partonic cross section of the incoming partons with hadron momenta fraction x_1, x_2. μ_R^2 is the scale at which the renormalization is performed, and μ_F^2 is called the "factorization scale", and is a parameter that separates the hard from the soft component. Both scales are typically chosen to be of the order of Q^2. The strong coupling and the PDFs are universal, and they can be measured in a wide variety of processes, in different experiments, and for different Q^2. Only σ_{ij} is process dependent and can be expanded in powers of α_s in pQCD.

1.3 Monte Carlo Simulation

MC simulation is an essential tool to mimic the complexity of hadron–hadron collisions. Even if the processes under study involve electromagnetic or weak interactions, the most challenging aspects of the MC simulation at hadron colliders are related to the description of the QCD phenomenology. MC generators provide cross section calculations in pQCD at a fixed order in α_s. Those are available at Leading Order (LO), and for some processes also at Next-to-LO (NLO). Some generators instead describe "multi-leg" processes (with multiple final state partons) with LO matrix elements for each parton multiplicity (see below). In order to take into account higher order corrections, MC simulations use the pQCD predictions at a given order supplemented by parton shower. In addition, the MC simulation include non-perturbative effects such as hadronization (formation of hadrons from partons) and underlying event (soft interaction between the remnant partons of the colliding hadrons). Figure 1.3 shows a sketch of a proton–proton collision as it is described in MC simulations.

The rest of this section introduces some key aspects of the MC simulation, and it presents the main features of the MC generators that are used in the various physics analyses in this manuscript.

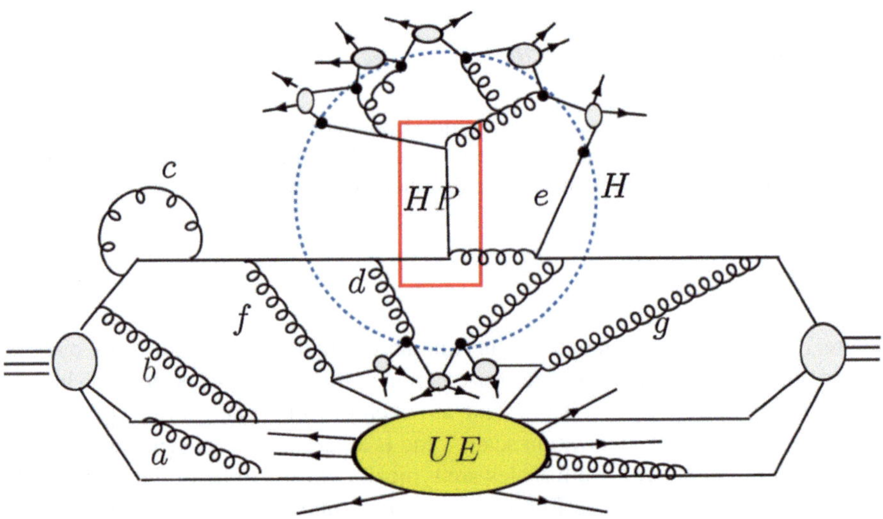

Fig. 1.3 Sketch MC description of a hadron–hadron collision. *HP* hard process, *UE* underlying event, and *H* factorization scale. The figure is taken from Ref. [13]

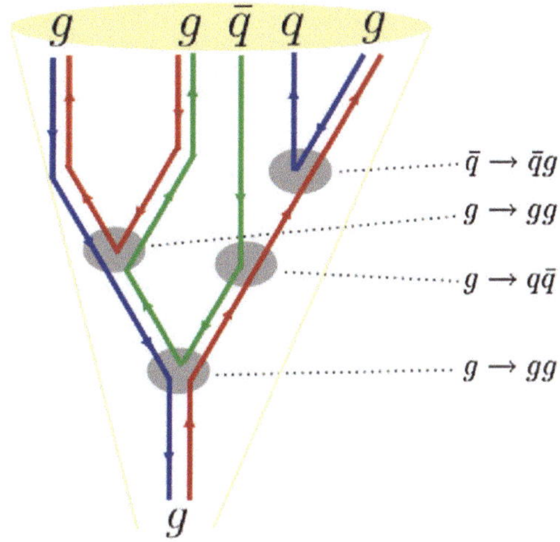

Fig. 1.4 Sketch of parton shower evolution

1.3.1 Parton Shower

Parton shower is the successive emission of quarks and gluons from the partons in the final (or initial) state. The emission is described by the DGLAP splitting functions calculated at LO, and in the limit of small angle emission. Sudakov form factors [14] are used to describe the probability that a parton evolves from an initial scale t_0 to a final scale t without radiating or splitting. A parton emitted in the final state is highly virtual, and "loses" virtuality at each radiation or branching. Subsequent branching takes place respecting the color connections (see Fig. 1.4). The procedure continues until the virtual mass of the parton reaches a cutoff scale Λ_{QCD}. On the initial-state radiation the algorithm is applied backwards in time. The initial parton decreases its virtuality and increases the carried fraction of momentum until it matches the PDFs x fraction. The outcome of successive branching is that the parton emitted in the hard scattering gives place to several partons that are typically collimated in a cone around the direction of the original parton.

1.3.2 Hadronization

After parton shower, colored partons recombine into final state color-neutral hadrons in a process called hadronization. The parton–hadron duality hypothesis assumes that the hadronization does not involve high momentum transfer, since partons combine with other partons close in phase space. This hypothesis is based on the observation that perturbation theory works properly down to a low scale Q $\sim$ 1 GeV. Therefore, hadronization does not change the original parton kinematics and flavor information, and the fundamental energy flow of the event is maintained. After parton shower and

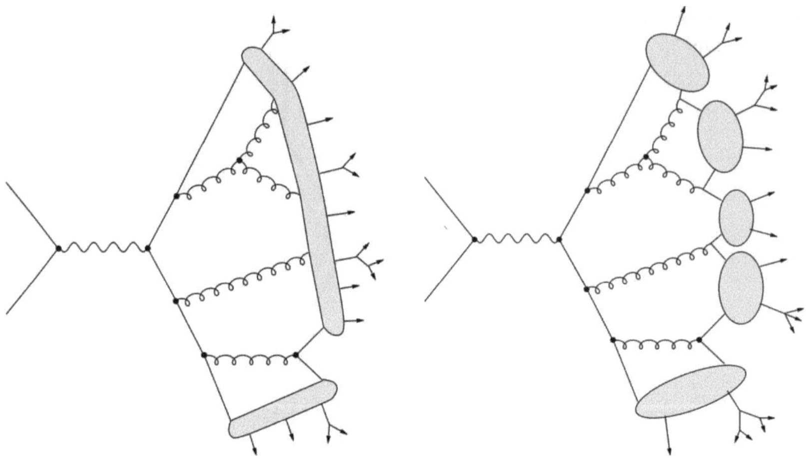

Fig. 1.5 Parton shower with hadronization from the string (*left*) or cluster (*right*) models

hadronization, jets can be defined from the final hadrons aiming to reproduce the kinematics of the original hard parton. Jets defined with particles after the hadronization are called "particle level" jets.

Parton–hadron duality is an observed property of QCD but it provides no details on how the hadronization takes place. Several phenomenological models are proposed, among which the string model and the cluster model are the most used (see Fig. 1.5). The string model uses string dynamics to describe the color flux between quarks. It assumes that the string between two quarks produces a linear confinement potential. If the quarks separate from each other, when the string energy exceeds the mass of a quark pair, the quark pair is created and the string is broken into two "shorter" strings of color-singlet states. The cluster model is based on the color pre-confinement of the branching processes. Gluons that remain after parton shower are split into quark-antiquark pairs. Neighboring quarks and antiquarks (not from the same gluon) are grouped in color-singlet clusters which typically decay into two hadrons. During the last decades, different measurements at colliders have been used to tune these models to properly describe the hadron multiplicity in the final state.

1.3.3 Underlying Event

In hadron–hadron colliders the partons that do not take part in the hard interaction can interact among them and contribute to the final state. Their interaction happens at low transfered momentum and involves flavor and color connections to the hard scattering, therefore it cannot be described perturbatively. Since the dominant QCD cross sections decrease rapidly with the transverse momentum of the hard scattering $\hat{p_{\mathrm{T}}}$, the underlying event is modeled as minimum bias events with low $\hat{p_{\mathrm{T}}} > \hat{p_{\mathrm{T}}}_{min}$. The parameter $\hat{p_T}_{min}$ has to be tuned from experimental data, using sensitive observables such as jet shapes or event energy flow.

1.3.4 Monte Carlo Generators

The main features of the MC generators used in the next chapters are described in the following. Emphasis is given to the aspects that are more relevant for the analyses treated in this thesis.

1.3.4.1 General Purpose Monte Carlo Generators: PYTHIA and HERWIG

PYTHIA [15] and HERWIG [16] are general purpose MC event generators, that use matrix elements at LO, and that include the simulation of both hard and soft interaction. These generators are specialized in the detailed description of parton showers in 2-to-1 and 2-to-2 processes. For the hadronization, HERWIG uses the cluster model, while the string model is used in PYTHIA. For the underlying event, HERWIG is typically interfaced with JIMMY MC [17]. Both PYTHIA and JIMMY simulate the underlying event as a scattering between proton remnants using 2-to-2 matrix elements at LO.

1.3.4.2 ALPGEN Monte Carlo

ALPGEN [18] is a MC event generator specialized on multi-parton hard processes at hadron–hadron collisions. In this thesis, ALPGEN is used to simulated Z+jets and W+jets events, but it can be used to generate other processes, such as $t\bar{t}$+jets production. In ALPGEN, events are generated with different multiplicities of outgoing partons, and cross sections are calculated at LO, for each parton multiplicity, using the ALPHA algorithm [19]. For the parton shower and hadronization, ALPGEN is interfaced with another MC program (such as PYTHIA or HERWIG) because ALPGEN only provides generation of events at parton-level.

The addition of parton shower introduces a double-counting of events. This is because the effect of parton shower on a sample with n-partons can produce additional jets that are already taken into account in the $n + 1$-partons sample. For this reason, ALPGEN uses the MLM matching technique, to separate the phase space simulated with matrix elements and that simulated with parton shower. This technique consists of three steps. At first the events are generated in samples of different parton multiplicities $n = 1, 2, 3 \ldots k$. Second, the parton shower development and the hadronization are employed separately for each sample. In the third step, a jet algorithm (typically cone-algorithm with radius R_{jet}) is run over the final state particles, and the resulting jets are matched with the partons from the matrix element calculation. The matching is based on the distance in η-ϕ-space, between the parton and the jet, typically $\Delta R = \sqrt{\Delta\eta^2 + \Delta\phi^2} < R_{jet}$. An important parameter of this procedure is the minimum $p_{\rm T}$ at which jets are defined, that is aften called "jet matching scale". If all partons are matched and if there is no additional jet ($N_{jet} = n$) the event is accepted, otherwise is rejected. For the sample with the highest parton multiplicity the number of jets can exceed the number of partons ($N_{jet} \geq k$), to take into account the radiation of additional partons that is not simulated with matrix elements in other event samples.

Finally, the samples with different parton multiplicities are normalized to their corresponding LO cross section, and are combined forming an inclusive sample. Subsequently, this inclusive sample is typically normalized to an inclusive cross section calculated at higher order in pQCD. At present, the inclusive production of Z and W bosons at hadron colliders is known at NNLO level using programs, such as MCFM [20] or FEWZ [21].

1.3.4.3 MC@NLO Monte Carlo

MC@NLO [22] produces hard scattering processes at NLO. Its generated events are typically used as input to HERWIG for the parton shower and hadronization, and to JIMMY for the underlying event. The use of 1-loop corrections leads to the generation of positive and negative weights for each event. Including full NLO corrections to the matrix elements reduced theoretical uncertainties on the cross section of inclusive processes. At the same time, the disadvantage of this approach is that the higher multiplicity parton emission relies on parton shower, that has a poorer description of well-separated hard radiation.

1.3.4.4 SHERPA Monte Carlo

SHERPA [23] is a multi-purpose event generator, that is interfaced with PYTHIA for the parton shower, and uses a multiple parton scattering model for underlying event simulation. The emission of multiple hard partons is handled with tree-level LO matrix elements, as it is done in ALPGEN. In order to separate the matrix element and parton shower domains, SHERPA uses the CKKW procedure [24]. This technique uses the k_T algorithm to generate a parton branching history. The matrix element is re-weighted using the value of α_s in every vertex of the branching, and the Sudakov factor in every line between vertices. Subsequently, hard emission that give place to a separate jet are vetoed.

SHERPA's modular design aims for a simple implementation of several processes and new MC techniques. For example, it is possible to introduce NLO corrections in the CKKW matching scheme using a NLO MC generator. This is done using the MENLOPS procedure [25], that corrects the inclusive cross section to NLO using POWHEG [26] MC generator, and simulate hardest emissions with tree-level matrix elements at LO.

1.4 Models for Physics Beyond the Standard Model

The SM is the most successful theory to describe leptons, quarks and their interactions through the strong and electroweak interactions. The predictions of this theory have been tested at high energy physics experiments for the last decades. Despite its success within its domain, the SM is not believed to be the fundamental theory to describe nature for a number of limitations.

One of the most discussed issues of the SM is called "hierarchy problem", and is related to the presence in the theory of the Higgs scalar field. In the SM, there is no mechanism to prevent scalar particles from acquiring large masses through radiative corrections. Therefore, the Higgs mass m_H receives enormous quantum corrections from every particle which couples to the Higgs field. Being m_H^0 the bare mass of the Higgs boson and Δm_H the radiative corrections to the Higgs mass

$$m_H = (m_H^0)^2 + \Delta m_H^2. \tag{1.7}$$

The corrections Δm_H^2 can be written as

$$\Delta m_H^2 = -\frac{\lambda_f^2}{16\pi^2}\left(2\Lambda^2 + \mathcal{O}\left[m_f^2\, ln\left(\frac{\Lambda}{m_f}\right)\right]\right), \tag{1.8}$$

where λ_f and m_f are the Yukawa couplings and the masses of the fermions, and Λ is an energy cutoff which is interpreted as the energy scale up to which SM is valid. For a given scale Λ, one can always choose a value of m_H^0 such that the observable mass m_H will be in agreement with measurements. However, if the SM needs to describe nature until the Planck scale, the quantum correction Δm_H^2 would be 30 orders of magnitude larger than m_H^2. Without an automatic cancellation of these corrections, this phenomenon would need a large "fine tuning" that is considered highly unnatural.

The SM does not include a description of the gravitational force, for which a quantization of general relativity is needed. Therefore, the SM cannot be a satisfactory "theory of everything".[2] It is expected that a new theory will be required at the energy scale where quantum gravitational effects become important. This scale is called Planck scale $M_{pl} = 1.22 \times 10^{19}$ GeV.

Another important drawback of the SM is related to the nature of the "Dark Matter" (DM), whose existence can be inferred by several cosmological observations, such as measure of the Cosmic Microwave Background and the rotational speed of the galaxies, among others. A more detailed discussion on DM is given in Sect. 1.4.2.

These and other fundamental problems motivate the idea that the SM is only an effective model of a more general theory. In the following sections, three scenarios for physics beyond the SM will be reviewed. These models are of particular interests

---

[2]This is usually the name of the hypothetical theory that would unify gravity and the other three forces.

in this thesis, because they predict new observable phenomena in the energy reach of the LHC and that would appear with a mono-jet signature.

1.4.1 Graviton Production in the ADD Scenario

Several extensions of the SM include the idea of extra-dimensions. In the model proposed by Arkani-Hamed, Dimopoulos, and Dvali [27] (ADD), new compact spatial extra-dimensions are added to the four dimensional time-space. This model assumes also that gravity can propagate in the higher-dimensional space, while SM particles are confined to the usual four dimensions. These assumptions result in an apparent weakness of gravity at large distances, while at small distances it can be as strong as the other interactions.

In the following, we will assume n compact extra-dimensions with the same radius R. The gravitation potential of two masses m_1 and m_2, placed at distance $r \ll R$ is

$$V(r) \sim \frac{m_1 m_2}{M_D^{n+2}} \frac{1}{r^{n+1}} \quad (\text{for } r \ll R), \tag{1.9}$$

where M_D^{n+2} is the real scale of gravity in the $4+n$ dimensions. If the masses are placed at distances $r \gg R$, their gravitational flux lines cannot continue to penetrate in the extra-dimensions, and the usual $1/r$ potential is obtained,

$$V(r) \sim \frac{m_1 m_2}{M_D^{n+2}} \frac{1}{R^n r} \quad (\text{for } r \gg R). \tag{1.10}$$

Therefore, the gravity scale M_D is related with the Planck mass M_{Pl} through the relation

$$M_{Pl}^2 \sim M_D^{n+2} R^n. \tag{1.11}$$

Depending on R, the gravity scale M_D could be as low as the electroweak scale (M_{EW}). This case is of particular interest since if these two scales are close to each other the hierarchy problem would be overcome, or at least reduced. In this picture, it does not appear implausible that gravity effects can start being measurable at energies much lower than M_{Pl}, possibly as low as M_{EW}. This has the exciting implication that high-energy colliders, as the LHC, can directly probe the physics of quantum-gravity.

For the following, $M_D \sim M_{EW}$ will be assumed, resulting in a prediction of the typical radius of

$$R \sim 10^{\frac{30}{n}-17}\text{cm} \left(\frac{1\ \text{TeV}}{M_{EW}}\right)^{1+\frac{2}{n}}. \tag{1.12}$$

The possibility of $n = 1$ is excluded by experimental evidences, because it would result in a $R \sim 10^{13}$ cm causing deviations of the gravity in the range of the solar system distances. The $n > 1$ instead, cannot be excluded by astrophysical observations or by measurements of the gravitational interactions at small distances (<1 mm).

Considering the scenario proposed by ADD, a low energy effective field theory is used to describe the infrared behavior of the gravitational interaction with SM particles [28]. The only field that respects the Einstein equation in $4+n$ dimensions and that describes a physical particle is called the "graviton", and is a spin-two object that mediates of the gravitation interaction. In the framework of compact extra-dimensions, gravitons result as a sum of different modes (called Kaluza-Klein, or KK modes) each with its own mass m_i. This is a general feature of theories with compact extra-dimensions, that were already introduced by Kaluza and Klein [29, 30] in 1921 and 1926, respectively. The Einstein equation in $4+n$ dimensions also establishes the Lagrangian of the free graviton field $G^{(k)}_{\mu\nu}$, and it's interaction term, that reads:

$$\mathcal{L}_{grav} = -\frac{1}{\bar{M}_{pl}} G^{(k)}_{\mu\nu} T^{\mu\nu}, \quad (1.13)$$

where $\bar{M}_{pl} = M_{pl}/\sqrt{8\pi}$ is the reduced Planck mass, and $T^{\mu\nu}$ is the energy-momentum tensor. Expanding $T^{\mu\nu}$ in Eq. 1.13, one retrieves the Feynman rules for graviton fields interacting with SM fields. Figure 1.6 shows the leading diagrams relevant for graviton production in association with partons.

In the case of $n \lesssim 6$, the mass difference between the graviton modes is small and the contributions of the different modes can be integrated over the mass. The differential cross section can be written as:

$$\frac{d^2\sigma}{dtdm} = \frac{2\pi^{n/2}}{\Gamma(n/2)} \frac{\bar{M}^2_{pl}}{M_D^{n+2}} m^{(n-1)} \frac{d\sigma_m}{dt}, \quad (1.14)$$

where $d\sigma_m/dt$ is the differential cross section of the KK mode with mass m, and t is the Mandelstam variable. For the different production $d\sigma_m/dt$ has the form:

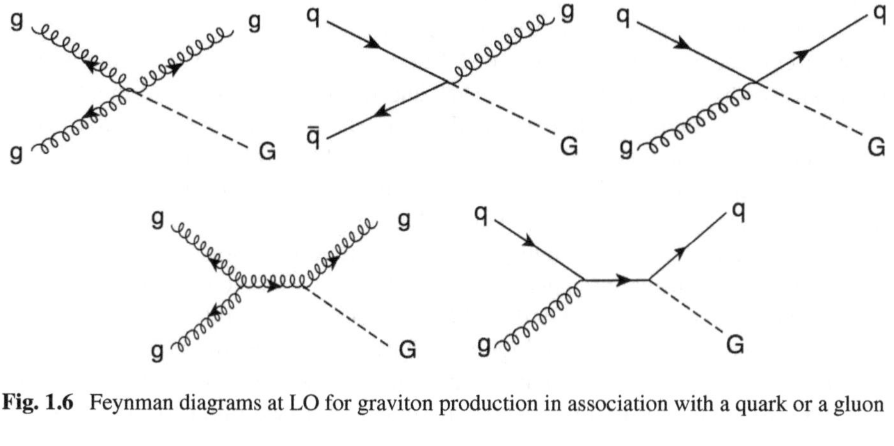

Fig. 1.6 Feynman diagrams at LO for graviton production in association with a quark or a gluon

$$\frac{d\sigma_m}{dt}(q\bar{q} \to gG) = \frac{\alpha_s}{36} \frac{1}{s\bar{M}_{pl}^2} F_1(t/s, m^2/s), \tag{1.15}$$

$$\frac{d\sigma_m}{dt}(qg \to qG) = \frac{\alpha_s}{96} \frac{1}{s\bar{M}_{pl}^2} F_2(t/s, m^2/s), \tag{1.16}$$

$$\frac{d\sigma_m}{dt}(gg \to gG) = \frac{3\alpha_s}{16} \frac{1}{s\bar{M}_{pl}^2} F_3(t/s, m^2/s), \tag{1.17}$$

where the expression of F_1, F_2, and F_3 are reported in Ref. [28].

An important remark is related to the ultraviolet validity cutoff of the effective field theory. This is an important issue because this cutoff determines the highest energy at which the predictions can be trusted. Unfortunately, a precise choice of the cutoff would strongly depend on the underlying fundamental theory, that is unknown. As mentioned before, the effect of the hypothetical underlying theory are expected to emerge at energy scales close to M_D, and therefore this is chosen as the cutoff of the theory.

1.4.2 Dark Matter and WIMP Pair Production at LHC

There is a certain consensus among physicists about the existence in the universe of a non-luminous matter, called "dark matter" (DM), that interacts gravitationally with SM particles (see Ref. [31] for a recent review). One of the most striking evidence of DM, comes from the measurement of the rotational velocity of stars in spiral galaxies. These galaxies are composed of a disc of matter that rotates around an axis, and the measurement of the rotational velocity as a function of the radius gives information about the amount of matter and its distribution inside the galaxy. Observations made until now require an amount of matter much larger than expected (see Fig. 1.7), and are compatible with the presence of a large DM halo, with a mass three to ten times larger than that corresponding to the visible matter.

Other evidencies of the existence of DM come from observations of the motion of galaxies within galaxy clusters [33, 34], and from measurements of the "cosmic microwave background" (CMB). The CMB is the thermal radiation background that is measured in the universe, corresponding to roughly 2.7 K of temperature. It is almost homogeneous in all directions (in a part over 10^5), and is stronger in the microwave sector of the spectrum. This radiation is not associated with a specific object (star, galaxy…) and is interpreted as the relic radiation emitted when the universe became transparent to photons (3×10^5 years after the Big Bang). Detailed study of the angular correlation in the CMB fluctuations gives information about the geometry of the universe (confirming that is flat), about its evolution in early stages, as well as its energy-matter content. At present, the most precise measurements of the CMB is made with the PLANCK experiment [35] (see Fig. 1.8). As measured

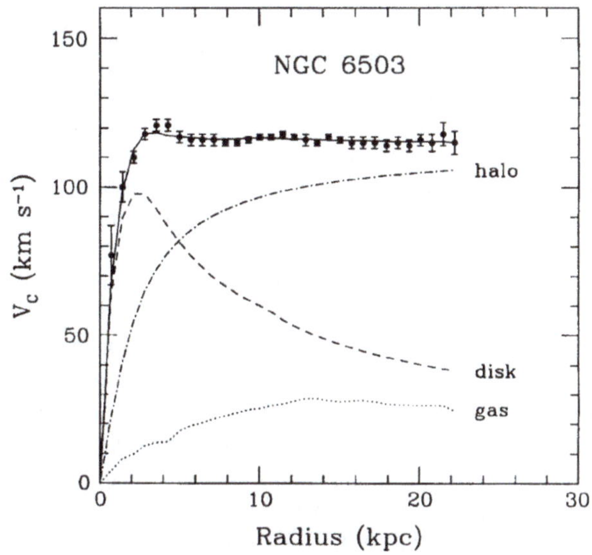

Fig. 1.7 Rotational velocity of stars as a function of the radius in the spiral galaxy NGC6503 [32]. The contributions of the disk (*dashed curve*), gas (*dotted curve*), and dark matter halo (*dot-dashed curve*) are also shown separately

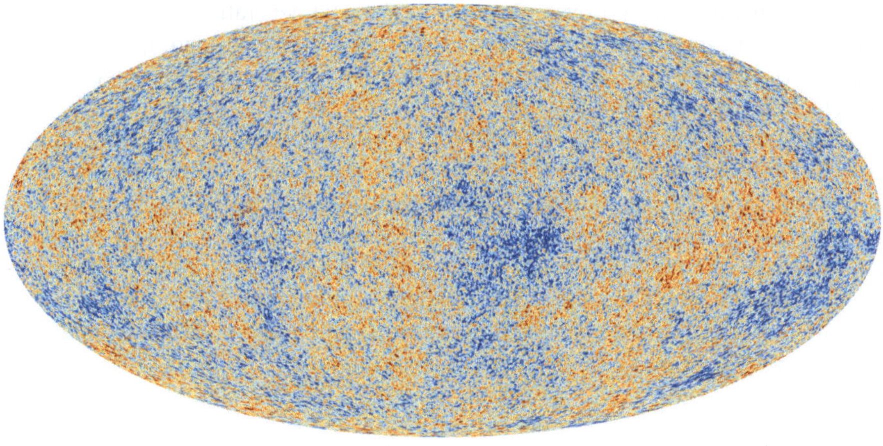

Fig. 1.8 Cosmic Microwave Background measured by the PLANCK experiment [35]. The colors represent the variation of measured temperature. The *blue* and *red* areas differ from each other by a temperature of 0.6 mK

from the PLANCK data, in the present universe the density of the ordinary matter, DM, and dark energy are 5, 27, and 68 %, respectively.

From the CMB measurement, one can also infer that DM was present in much larger quantity in the early universe. A striking coincidence in cosmology is that if DM would annihilate to SM particles with an interaction strength close that of the weak force, that would result exactly in the decrease of DM density observed between early and present universe (see Ref. [36] for a concise review). This coincidence leads to the idea that DM could be composed of Weak Interacting Massive Particles (WIMPs). These are expected to have a mass between few GeV and a TeV and couple to SM particles through a generic weak interaction. Many new models for physics beyond the SM designed to solve the hierarchy problem (Supersymmetry for instance) also predict WIMP candidates.

WIMPs have being searched for with a variety of detection strategies. In "direct detection" experiments the aim is to observe WIMP-nucleon elastic scattering , by measuring the nuclear recoil. Instead, "indirect detection" experiments search for the SM products from WIMP annihilation. In the last decade, the field of DM detection has attracted a lot of interest because there have been several published results that can be interpreted as detection of WIMP particles. Possible hints of detection of a light WIMP ($\sim$10 GeV) have emerged from data obtained by the DAMA/LIBRA [37], CDMS II [38], CRESST-II [39], CoGeNT [40, 41], but the interpretation of these events as due to scattering of a WIMP has been challenged by several other experiments such as XENON100 [42, 43] (see Fig. 1.9). Indirect detection experiments, such as AMS [44], FERMI [45] and PAMELA [46], have shown an anomalous excess of high-momentum positrons in the galaxy. Such an excess is consistent with the hypothesis of WIMP annihilation, but not yet sufficiently conclusive to rule out other explanations, like for example pulsars. The next generation of direct and indirect detection experiments, characterized by very low background, larger volumes and improved energy resolution, is awaited to shine a light on these ambiguous hints of detection.

A third strategy to search for WIMPs is based on the direct production at colliders. In collider experiments, the WIMPs are undetected and if produced in association with other objects (such as jets or gauge bosons), result in final states with large missing energy. As it will be shown in this thesis, the sensibility of collider searches is competitive with (and in some cases higher than) those of direct and indirect detection experiments. In case of discovery, the complementarity of these three strategies (see sketch in Fig. 1.10) will be essential to measure the properties of the DM candidate(s), with less assumptions, and with reduced ambiguities.

1.4.2.1 WIMP Production at Hadron Colliders

Production of WIMPs at colliders can happen in many ways depending on the physics beyond the SM that one is considering. In the following, an effective field theory is used to describe possible interactions between WIMPs and partons (see Ref. [50]). Here, WIMPs are assumed to be a Dirac-like fermion, and odd under the Z_2 symmetry (R-parity in SUSY, or KK-parity in extra dimensions), so that each coupling involves an even number of WIMPs. Different effective operators (listed in Table 3.10) are taken into account to mimic the different nature of the mediators in a Fermi-like point interaction. For each operator, a parameter M^* characterizes the strength of the

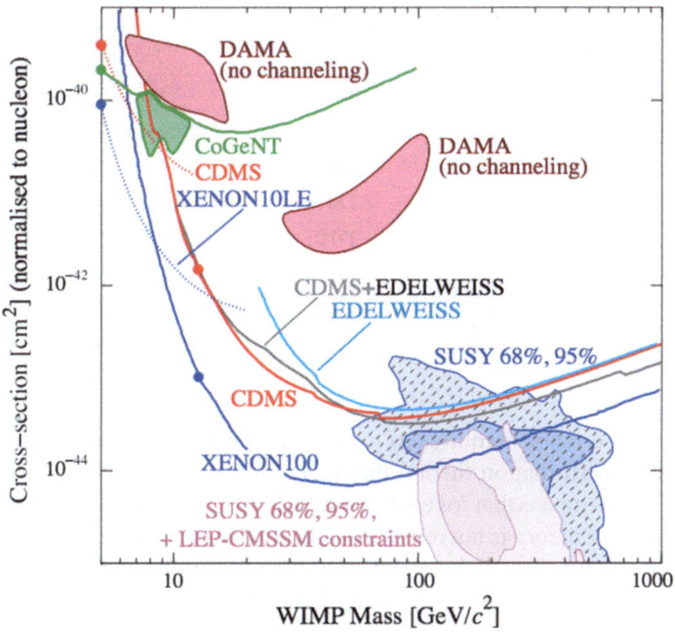

Fig. 1.9 Cross sections for spin-independent coupling versus WIMP mass. The figure is taken from Ref. [47], that details the experimental results. *Shaded* 68 and 95 % CL regions are from Supersymmetry predictions on WIMP candidates, that include recent LHC constraints, and are taken from Refs. [48, 49]

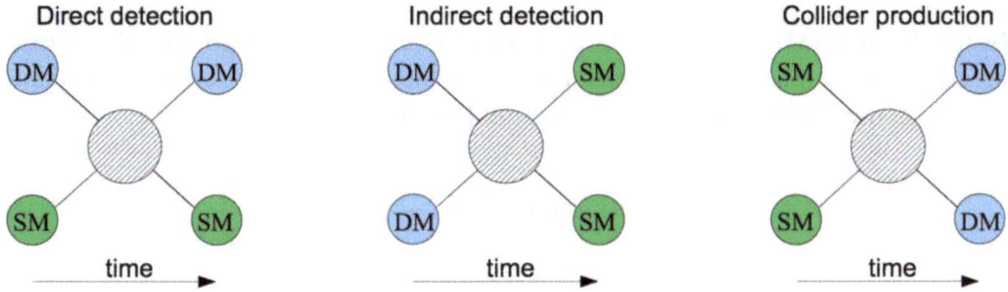

Fig. 1.10 Sketch of the different strategies for Dark Matter search

interaction. In Table 3.10, the operators D1 and D11 correspond to a scalar mediator that couples to quarks and gluons, respectively. The operators D5 and D8 would correspond instead to a vector and axial vector mediator, respectively.

The effective field theory used here is a good approximation of the real theory if the mass of the mediator (or mediators) is too heavy to be produced directly. Supposing a mediator of mass M, the suppression scale $M^* \sim M(g_{SM}g_{DM})^{-\frac{1}{2}}$ where g_{SM} and g_{DM} are the coupling of the mediator with SM and DM particles. So even for moderate M^* the theory can be still valide if the couplings are sufficiently large.

Table 1.2 Operators coupling Dirac fermion WIMPs to Standard Model quarks or gluons

Name	Operator	Name	Operator
D1	$\frac{m_q}{(M^*)^3}\bar{\chi}\chi\bar{q}q$	D2	$\frac{m_q}{(M^*)^3}\bar{\chi}\gamma^5\chi\bar{q}q$
D3	$\frac{m_q}{(M^*)^3}\bar{\chi}\chi\bar{q}\gamma^5 q$	D4	$\frac{m_q}{(M^*)^3}\bar{\chi}\gamma^5\chi\bar{q}\gamma^5 q$
D5	$\frac{1}{(M^*)^2}\bar{\chi}\gamma^\mu\chi\bar{q}\gamma_\mu q$	D6	$\frac{1}{(M^*)^2}\bar{\chi}\gamma^\mu\gamma^5\chi\bar{q}\gamma_\mu q$
D7	$\frac{1}{(M^*)^2}\bar{\chi}\gamma^\mu\chi\bar{q}\gamma_\mu\gamma^5 q$	D8	$\frac{1}{(M^*)^2}\bar{\chi}\gamma^\mu\gamma^5\chi\bar{q}\gamma_\mu\gamma^5 q$
D9	$\frac{1}{(M^*)^2}\bar{\chi}\sigma^{\mu\nu}\chi\bar{q}\sigma_{\mu\nu}q$	D10	$\frac{1}{(M^*)^2}\epsilon^{\mu\nu\alpha\beta}\bar{\chi}\sigma^{\mu\nu}\gamma^5\chi\bar{q}\sigma_{\alpha\beta}q$
D11	$\frac{1}{(4M^*)^3}\bar{\chi}\chi\alpha_s(G^a_{\mu\nu})^2$	D12	$\frac{1}{(4M^*)^3}\bar{\chi}\gamma^5\chi\alpha_s(G^a_{\mu\nu})^2$
D13	$\frac{1}{(4M^*)^3}\bar{\chi}\chi\alpha_s G^a_{\mu\nu}\tilde{G}^{a,\mu\nu}$	D14	$\frac{1}{(4M^*)^3}\bar{\chi}\gamma^5\chi\alpha_s G^a_{\mu\nu}\tilde{G}^{a,\mu\nu}$

Since an effective theory requires $M > 2m_\chi$ and couplings are $g_{SM}g_{DM} \leq (4\pi)^2$ to be treated in perturbation theory, the interaction with strongest coupling satisfy $m_\chi = 2\pi M^*$. This means that for each m_χ there is a M^* lower bound of validity below which the effective theory is not reliable anymore. In case the effective approach does not strictly apply, it is hard to know whether the predictions under-estimate or over-estimate the cross sections due to the lack of knowledge on the underlying ultraviolet theory.

1.4.2.2 Collider Results Compared with Direct and Indirect Detection Experiments

Results from WIMP pair production at colliders can be compared with those from direct and indirect detection experiments. Exclusion limits on M^* are translated into bounds on the WIMP-nucleon cross section [50, 51]:

$$\sigma_0^{D1} = 1.60 \times 10^{-37}\text{cm}^2 \left(\frac{\mu_\chi}{1\text{ GeV}}\right)^2 \left(\frac{20\text{ GeV}}{M^*}\right)^6, \tag{1.18}$$

$$\sigma_0^{D5} = 1.38 \times 10^{-37}\text{cm}^2 \left(\frac{\mu_\chi}{1\text{ GeV}}\right)^2 \left(\frac{300\text{ GeV}}{M^*}\right)^4, \tag{1.19}$$

$$\sigma_0^{D8} = \sigma_0^{D9} = 9.18 \times 10^{-40}\text{cm}^2 \left(\frac{\mu_\chi}{1\text{ GeV}}\right)^2 \left(\frac{300\text{ GeV}}{M^*}\right)^4, \tag{1.20}$$

$$\sigma_0^{D11} = 3.83 \times 10^{-41}\text{cm}^2 \left(\frac{\mu_\chi}{1\text{ GeV}}\right)^2 \left(\frac{100\text{ GeV}}{M^*}\right)^6. \tag{1.21}$$

where μ_χ the reduced mass of the WIMP-nucleon system, $\mu_\chi = m_\chi * m_N/(m_\chi + m_N)$.

For this comparison, one needs to keep in mind the different kinematic regime in which the interaction is happening. In direct detection experiments, the typical transferred momentum is of the order of a keV. In this regime the propagator of a particle with mass $m \gg 1$ KeV that mediate the interaction cannot be resolved, making a Fermi-like point interaction suitable. On the contrary, at LHC the center of mass scale of the hard scattering $\sqrt{\hat{s}}$ can be up to the TeV scale.

Finally, results at colliders can be translated into WIMP annihilation cross section, relevant for indirect detection experiments. DM annihilation rate is proportional to the quantity $\langle \sigma v \rangle$, where σ is the annihilation cross section, v is the relative velocity of the annihilating particles, and the average is over the DM velocity distribution. Using the effective field theory approach already introduced, we find a σv for operator D5 (vector interaction), and for operator D8 (axial-vector interaction) are [52]:

$$\sigma^{D5} v = \frac{1}{16\pi M^{*4}} \sum_q \sqrt{1 - \frac{m_q^2}{m_\chi^2}} \left(24(2m_\chi^2 + m_q^2) + \frac{8m_\chi^4 - 4m_\chi^2 m_q^2 + 5m_q^4}{m_\chi^2 - m_q^2} v^2 \right), \quad (1.22)$$

$$\sigma^{D8} v = \frac{1}{16\pi M^{*4}} \sum_q \sqrt{1 - \frac{m_q^2}{m_\chi^2}} \left(24m_q^2 + \frac{8m_\chi^4 - 22m_\chi^2 m_q^2 + 17m_q^4}{m_\chi^2 - m_q^2} v^2 \right) \quad (1.23)$$

where the sum $\sum_q$ runs over all kinematically accessible quarks. Note that for other effective operators (for instance the scalar operator D1), the bounds from colliders can be much stronger, especially for low $\langle v^2 \rangle$. It has to be noticed that the annihilation rate $\langle \sigma v \rangle$ is calculated in the assumption of equal coupling for all quarks, and only annihilation to quarks is considered. In the case of DM coupled with leptons, the exclusion limits are weakened by a factor $1/\mathrm{BR}(\chi\chi \to q\bar{q})$.

1.4.3 Gravitino Production in the GMSB Scenario

Supersymmetry (SUSY) is the most popular among the scenarios beyond the SM. For each SM fermion this theory predicts a bosonic "super-partner", and viceversa for the SM bosons. The introduction of this symmetry is particularly appealing for solving several issues of the SM. First of all, SUSY eliminates the hierarchy problem because divergent corrections to the Higgs mass from fermions and their bosonic super-partner are automatically cancelled out (see Fig. 1.11). Additionally, this theory includes valid DM candidates and provides a framework in which the unification of gravity and the SM interactions can be more easily pursued.

SUSY needs to be a broken symmetry since no SUSY particle has been observed until now, and the masses of the superpartners are therefore different from those of their SM partners. Such a symmetry breaking needs to happen at a relatively low energy scale (10–100 TeV) in order to still provide a solution for the hierarchy problem. Typically, it is assumed that a spontaneous symmetry breaking is induced

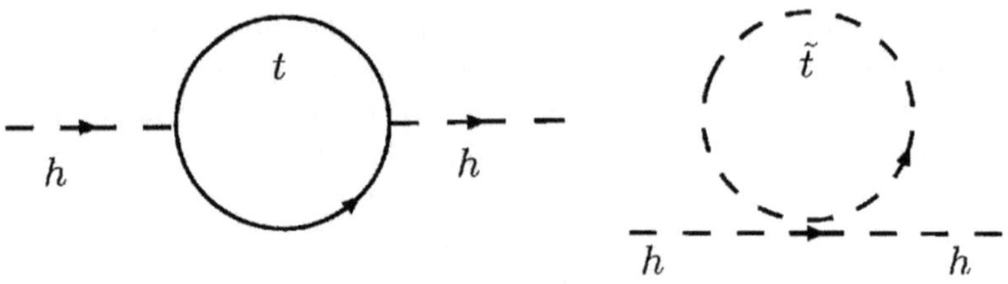

Fig. 1.11 Corrections to the Higgs self-energy with SUSY. The term with a top loop is compensated by a stop loop

by a hidden sector, and it is due to "soft breaking" terms added to the Lagrangian:

$$\mathcal{L} = \mathcal{L}_{SUSY} + \mathcal{L}_{soft}. \tag{1.24}$$

The breaking is communicated from the hidden to the visible sector either through gravity or electroweak and QCD gauge interactions. These two possible mechanisms for SUSY breaking lead to Minimal Supergravity (mSUGRA) [53, 54] and Gauge Mediated SUSY Breaking (GMSB) [55–57]. In the latter, messenger fields at mass scale M_{mess} are supposed to share the gauge interaction, and to provide the soft breaking terms in loop diagrams. The coupling of the messenger fields to the hidden sector produces in the fields a supersymmetric mass of order M with mass-squared splittings of order F, with $\sqrt{F}$ being the scale of SUSY breaking. In its minimal version, GMSB models are described by six fundamental parameters, that define the mass hierarchy of all particles, and therefore the phenomenology:

- $\sqrt{F}$: the scale of the SUSY breaking. SUSY masses are proportional to $\sqrt{F}$.
- M_{mess}: the mass scale of the messengers.
- N_5: number of messenger fields. Gaugino[3] masses depend linearly on N_5.
- tan(β): ratio of the two vacuum expectation values of the SUSY Higgs doublets.
- sign(μ): sign of the Higgsino mass term. Gaugino masses are dependent on this parameter.

In GMSB, the super-partner of the spin-2 graviton, the spin-3/2 gravitino $\tilde{G}$, is the lightest supersymmetric particle (LSP). The gravitino does not necessarily couple to matter with gravitational strength only, but its coupling can be enhanced to electroweak strength once SUSY is broken through the super-Higgs mechanism and the associated Goldstone fermion, the spin-1/2 goldstino, is absorbed to give the gravitino its mass. An important feature of this mechanism is that the gravitino mass gives direct access to the scale of the SUSY breaking:

$$\langle F \rangle = \sqrt{3}\, m_{\tilde{G}}\, \bar{M}_{pl}, \tag{1.25}$$

where $\bar{M}_{pl} = M_{pl}/\sqrt{8\pi}$.

[3]Gauginos are combinations of the SUSY electroweak and Higgs fermionic fields.

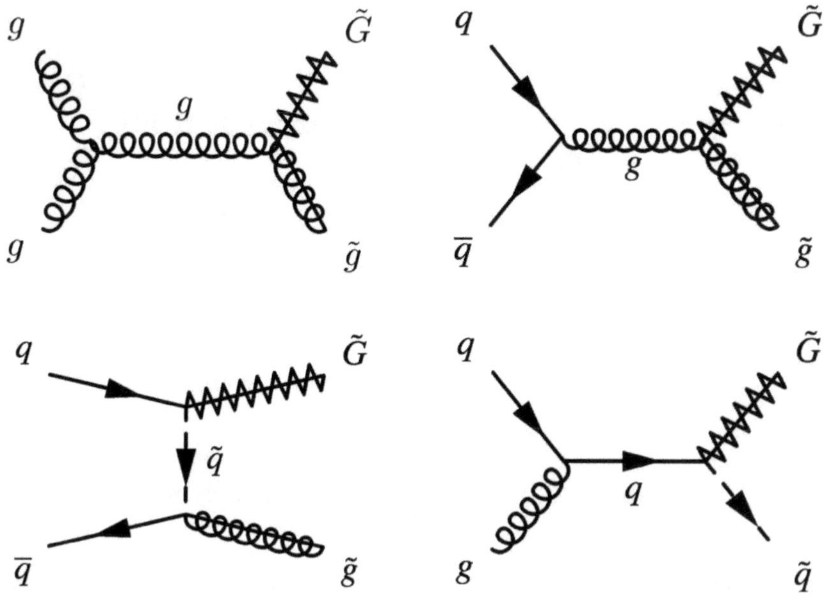

Fig. 1.12 Some of the LO diagrams for $\tilde{G}$ + $\tilde{q}/\tilde{g}$ production at LHC

In the following, the production of gravitinos in association with a squark or a gluino[4] at hadron colliders is described. Here a simplified scenario is considered, depending only on the gravitino, squark and gluino masses. In the case of very light gravitino, the productions (pp→ $\tilde{G}$+$\tilde{g}$) and (pp→ $\tilde{G}$+$\tilde{q}$) dominates over the strong production of squarks and gluinos, and the dominant squark and gluino decays are $\tilde{q} \to q\tilde{G}$ and $\tilde{g} \to g\tilde{G}$, respectively. Therefore, the final state is characterized by two gravitinos escaping detection and a jet, leading to a mono-jet final state.

The $\tilde{G}$+$\tilde{g}$ associated production is driven by two competing initial states, i.e. quark-antiquark or gluon–gluon scattering, while the production $\tilde{G}$+$\tilde{q}$ can only be produced in quark-gluon scattering (see Fig. 1.12). Predictions for $\tilde{G}$+$\tilde{g}/\tilde{q}$ are calculated at LO in pQCD, neglecting the gravitino mass everywhere apart in the coupling constants. The differential cross section, expressed in terms of the usual Mandelstam variables takes the form:

$$\frac{d\sigma}{dt} = \frac{1}{2s}\frac{1}{8\pi s}|\overline{M}|, \quad (1.26)$$

with

$$|\overline{M}|^2(gg \to \tilde{G}\tilde{g}) = \frac{g_s^2 m_{\tilde{g}}^2}{6C_F M_{pl}^2 m_{\tilde{G}}^2} F_A(u, t, s, m_{\tilde{g}}), \quad (1.27)$$

[4]The squark and the gluino are the superpartner of the squark and gluon, respectively.

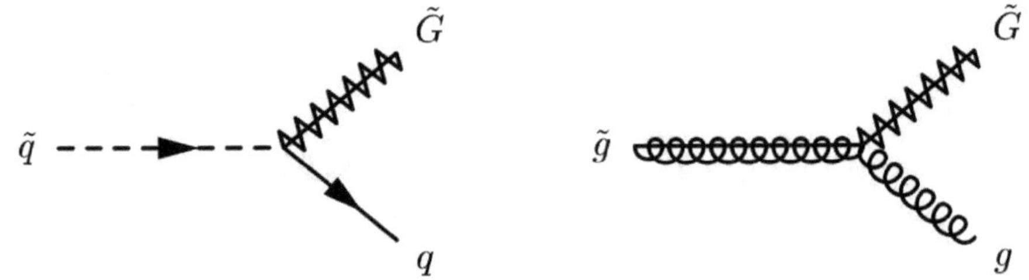

Fig. 1.13 LO diagrams for the decays $\tilde{q} \to q\tilde{G}$ (*left*) and $\tilde{g} \to g\tilde{G}$ (*right*)

$$|\overline{M}|^2(q\bar{q} \to \tilde{G}\tilde{g}) = \frac{g_s^2 C_F}{3N_C M_{pl}^2 m_{\tilde{G}}^2} F_B(u, t, s, m_{\tilde{g}}, m_{\tilde{q}}), \quad (1.28)$$

$$|\overline{M}|^2(qg \to \tilde{G}\tilde{q}) = \frac{g_s^2}{12N_C M_{pl}^2 m_{\tilde{G}}^2} F_C(u, t, s, m_{\tilde{q}}, m_{\tilde{g}}), \quad (1.29)$$

where the explicit expression of F_A, F_B, and F_C is reported in Ref. [58]. It can be noticed that the cross sections depend $m_{\tilde{G}}^2$ as $\sigma \sim 1/m_{\tilde{G}}^2$, and therefore lower bounds on $m_{\tilde{G}}$ can be deduced from the cross section constraints. It can also be noticed that the $\tilde{G}$+$\tilde{g}$ production has a dependency on $m_{\tilde{q}}$, coming from diagram contributions with squark exchange in the *t* channel. (see diagrams in Fig. 1.12). In the same way, the $\tilde{G}$+$\tilde{q}$ production cross section depends on the gluino mass $m_{\tilde{g}}$.

As previously mentioned, for a light gravitino the $\tilde{q} \to q\tilde{G}$ and $\tilde{g} \to g\tilde{G}$ decays dominate (Fig. 1.13). This assumption is studied in Ref. [58], showing that the branching ratios $BR(\tilde{q} \to \tilde{G}q)$ and $BR(\tilde{g} \to \tilde{G}g)$ are larger than 0.9, for gravitino masses $m_{\tilde{G}} \lesssim 10^{-4}$ GeV.

The branching ratios of squark or gluino decays are included in the cross sections using the Narrow Width Approximation (NWA). In this procedure, intermediate particles are set on shell ($\Gamma = 0$) in order to simplify the calculation. In our case, the width of squarks and gluinos are neglected so that the cross section can be factorized in the following way.

$$\sigma(pp \to \tilde{G}\tilde{G}q) \simeq \sigma(pp \to \tilde{G}\tilde{q}) \times BR(\tilde{q} \to \tilde{G}q) \ , \quad (1.30)$$

Typically, this approximation is considered valid if the width of the particle does not exceed 25 % of its mass.

References

1. S.L. Glashow, Partial symmetries of weak interactions. Nucl. Phys. **22**, 579–588 (1961)
2. S. Weinberg, A model of leptons. Phys. Rev. Lett. **19** (1967)

3. A. Salam, Gauge unification of fundamental forces. Rev. Mod. Phys. **52**, 525538 (1980)
4. J. Beringer et al., Particle data group. Phys. Rev. D **86**, 010001 (2012)
5. ATLAS Collaboration, Observation of a new particle in the search for the standard model higgs boson with the ATLAS detector at the LHC. Phys. Rev. D **716**, 1–29 (2012a)
6. C.M.S. Collaboration, Observation of a new boson at a mass of 125 GeV with the CMS experiment at the LHC. Phys. Rev. D **716**, 30–61 (2012b)
7. ATLAS Collaboration, Combined measurements of the mass and signal strength of the Higgs-like boson with the ATLAS detector using up to 25 fb^{-1} of proton-proton collision data. (ATLAS-CONF-2013-014) (2013a)
8. F. Englert, R. Brout, Broken symmetry and the mass of gauge vector mesons. Phys. Rev. Lett. **13**, 321 (1964)
9. P.W. Higgs, Broken symmetries and the masses of gauge bosons. Phys. Rev. Lett. **13**, 508 (1964)
10. C.R. Hagen, G.S. Guralnik, T.W.B. Kibble, Global conservation laws and massless particles. Phys. Rev. Lett. **13**, 585 (1964)
11. D.J. Gross, F. Wilczek, Asymptotically free Gauge Theories. Phys. Rev. D **8**, 3633 (1973)
12. S. Bethke, Eur. Phys. J. C **64**, 689 (2009)
13. T.J. Stelzer, M.L. Mangano, Tools for the simulation of hard hadronic collisions
14. V.V. Sudakov, Vertex parts at very high energies in quantum electrodynamics. Sov. Phys. JETP **3**, 65 (1956)
15. S. Mrenna, T. Sjostrand, P. Skands, PYTHIA 6.4 physics and manual. JHEP **05**, 026 (2006)
16. G. Corcella et al., HERWIG 6: An event generator for hadron emission reactions with interfering gluons (including supersymmetric processes). JHEP **01**, 010 (2001)
17. J. Forshaw, J. Butterworth, M. Seymour, Multiparton interactions in photoproduction at hera. Z. Phys., **C72**, 637–646 (1996)
18. M. Mangano et al., Alpgen, a generator for hard multiparton processes in hadronic collisions. JHEP **07**, 001 (2003)
19. F. Caravaglios, M.L. Mangano, M. Moretti, R. Pittau, A new approach to multi-jet calculations in hadron collisions. Nucl. Phys. B **539**, 215 (1999)
20. J.M. Campbell, R.K. Ellis, Phys. Rev. D **65**, 113007 (2002)
21. R. Gavin, Y. Li, F. Petriello, S. Quackenbush, FEWZ 2.0: A code for hadronic Z production at next-to- next-to-leading order (2010)
22. S. Frixione, B.R. Webber, The MC@NLO 3.2 event generator (2006)
23. F. Krauss, M. Schonherr, S. Schumann, et al., T. Gleisberg, S. Hoeche, Event generation with SHERPA 1.1. JHEP **007**, 0902 (2009)
24. R. Kuhn, B.R. Webber, S. Catani, F. Krauss, JHEP (2001)
25. P.N. Keith Hamilton, Improving nlo-parton shower matched simulations with higher order matrix elements (2010)
26. C. Oleari, S. Alioli, P. Nason, E. Re, A general framework for implementing nlo calculations in shower monte carlo programs: the powheg box. JHEP 1006 (2010)
27. N. Arkani-Hamed, S. Dimopoulos, G.R. Dvali, The hierarchy problem and new dimensions at a millimeter. Phys. Lett. B **429**, 263–272 (1998). doi:10.1016/S0370-2693(98)00466-3
28. G.F. Giudice, R. Rattazzi, J.D. Wells, Quantum gravity and extra dimensions at high-energy colliders (2000)
29. T. Kaluza. Zum Unittsproblem in der Physik. Sitzungsber. Preuss. Akad. Wiss. Berlin. (Math. Phys.) 966972 (1921)
30. O. Klein, Quantentheorie und fnfdimensionale Relativittstheorie. Zeitschrift fr Physik **A 37**, 895–906 (1926)
31. G. Bertone, D. Hooper, J. Silk, Particle dark matter: Evidence, candidates and constraints. Phys. Rept. **405**, 279–390 (2005). doi:10.1016/j.physrep.2004.08.031
32. A.H. Broeils, K.G. Begeman, R.H. Sanders, Extended rotation curves of spiral galaxies—dark haloes and modified dynamics. Monthly Not. R. Astron. Soc. **249**, 523 (1991)
33. F. Zwicky, On the masses of Nebulae and of clusters of Nebulae. Astrophys. J. 217 (1937)

34. M. Girardi, et al., Optical luminosities and mass-to-light ratios of nearby galaxy clusters. Astrophys. J. 62 (2000)
35. Planck Collaboration, Planck 2013 results. I. Overview of products and scientific results (2013b)
36. V. Zacek, D. Matter, Proceedings of the 2007 Lake Louise Winter Institute (2007)
37. F. Cappella, R. Cerulli, C. Dai, et al., R. Bernabei, P. Belli, New results from DAMA/LIBRA. Eur. Phys. J. **C67**, 39 (2010)
38. R. Agnese, et al., (CDMS Collaboration), Dark matter search results using the silicon detectors of CDMS II (2013)
39. I. Bavykina, A. Bento, C. Bucci, et al., G. Angloher, M. Bauer, Results from 730 kg days of the CRESST-II Dark Matter Search. Eur. Phys. J. **C72**, 1971 (2012)
40. J. Colaresi, J. Collar, J. Diaz Leon, et al., C. Aalseth, P. Barbeau, Search for an annual modulation in a P-type point contact germanium dark matter detector. Phys. Rev. Lett. 141301 (2011)
41. C. Aalseth et al., CoGeNT Collaboration. Results from a search for light-mass dark matter with a p-type point contact germanium detector. Phys. Rev. Lett. **106**, 131301 (2011)
42. XENON100 Collaboration, Dark matter results from 225 live days of xenon100 data
43. E. Aprile et al., Dark matter results from 100 live days of XENON100 data. Phys. Rev. Lett. **107**, 131302 (2011). doi:10.1103/PhysRevLett.107.131302
44. M. Aguilar, et al., (AMS Collaboration), First result from the alpha magnetic spectrometer on the international space station: Precision measurement of the positron fraction in primary cosmic rays of 0.5350 GeV. Phys. Rev. Lett. 141102 (2013)
45. M. Ackermann, et al., Constraining dark matter models from a combined analysis of milky way satellites with the fermi large area telescope. Phys. Rev. Lett. **107**, 241302 (2011). (6 pages, 2 figures/ Contact authors: Johann Cohen-Tanugi, Jan Conrad, and Maja Llena Garde)
46. PAMELA Collaboration, An anomalous positron abundance in cosmic rays with energies 1.5100 GeV. Nature **458**, 607–609 (2009)
47. J. Beringer, et al., (Particle Data Group). http://pdg.lbl.gov/2012/reviews/rpp2012-rev-dark-matter.pdf
48. R. Trotta et al., The impact of priors and observables on parameter inferences in the Constrained MSSM. JHEP **024**, 0812 (2008a)
49. O. Buchmueller et al., The CMSSM and NUHM1 in Light of 7 TeV LHC, Bs→ $\mu^+\mu^-$ and XENON100 Data. Eur. Phys. J. C **71**, 1634 (2011)
50. Jessica Goodman, Masahiro Ibe, Arvind Rajaraman, William Shepherd, Tim M.P. Tait et al., Constraints on dark matter from colliders. Phys. Rev. D **82**, 116010 (2010). doi:10.1103/PhysRevD.82.116010
51. A. Pukhov, G. Belanger, F. Boudjema, A. Semenov, Dark matter direct detection rate in a generic model with micrOMEGAs2.2. Comput. Phys. Commun. 747–767 (2008)
52. Patrick J. Fox, Roni Harnik, Joachim Kopp, Yuhsin Tsai, Missing energy signatures of dark matter at the LHC. Phys. Rev. D **85**, 056011 (2012). doi:10.1103/PhysRevD.85.056011. 22 pages, 10 figures
53. H.P. Nilles, Supersymmetry, supergravity and particle physics. Phys. Rep. **110**, 1 (1984)
54. R. Arnowitt, A.H. Chamseddine, P. Nath, Locally supersymmetric grand unification. Phys. Rep. **49**, 970 (1982)
55. G. Giudice, R. Rattazzi, Theories with gauge mediated supersymmetry breaking. Phys. Rept. **322**, 419–499 (1999)
56. P. Fayet, Mixing between gravitational and weak interactions through the massive gravitino. Phys. Lett. B **70**, 461 (1977)
57. D. Dominici, F. Feruglio, R. Casalbuoni, S. De Curtis, R. Gatto, A gravitino—goldstino high-energy equivalence theorem. Phys. Lett. B **215**, 313 (1988)
58. M. Klasen, G. Pignol, New results for light gravitinos at hadron colliders: Tevatron limits and LHC perspectives. Phys. Rev. D **75**, 115003 (2007)

Chapter 2
The ATLAS Experiment at LHC

This chapter introduces the main aspects of the ATLAS detector at the LHC collider. The reconstruction procedures of the physics objects that are relevant for the analyses described in this thesis (jets, electrons, muons and missing energy) are also discussed. The Appendix A describes dedicated studies carried out during the commissioning of the Tile hadronic calorimeter (TileCal), while Appendix B details the measurement of the instantaneous luminosity using the TileCal data.

2.1 Large Hadron Collider

The Large Hadron Collider (LHC) [1] is a particle accelerator designed to collide protons at a center of mass energy $\sqrt{s} = 14\,\text{TeV}$. On the accelerator ring ($\sim$27 km in circumference) four detectors (ALICE [2], ATLAS [3], CMS [4] and LHCb [5]) have been built around the four interaction points to reconstruct and study the collisions delivered by the LHC.

Since 2010, the LHC has delivered proton-proton (p-p) collisions at center of mass energies of 7 and 8 TeV, about half of its nominal energy. The LHC has produced also lead ions (Pb-Pb) collisions with a per-nucleon center of mass energy $\sqrt{s_{NN}} = 2.76\,\text{TeV}$ and proton-ion (p-Pb) collisions with $\sqrt{s_{NN}} = 5.02\,\text{TeV}$. More details about the delivered luminosity will be given below.

2.2 The ATLAS Detector

ATLAS (A Toroidal LHC ApparatuS) [3] is a general purpose detector designed for the reconstruction and identification of jets, photons, electrons, muons, taus and missing transverse energy. The detector has been designed to optimize the search for the Higgs boson and for a large number of searches for new phenomena.

V. Rossetti, *Search for Exotic Mono-jet Events*, Springer Theses,
DOI 10.1007/978-3-319-22225-7_2

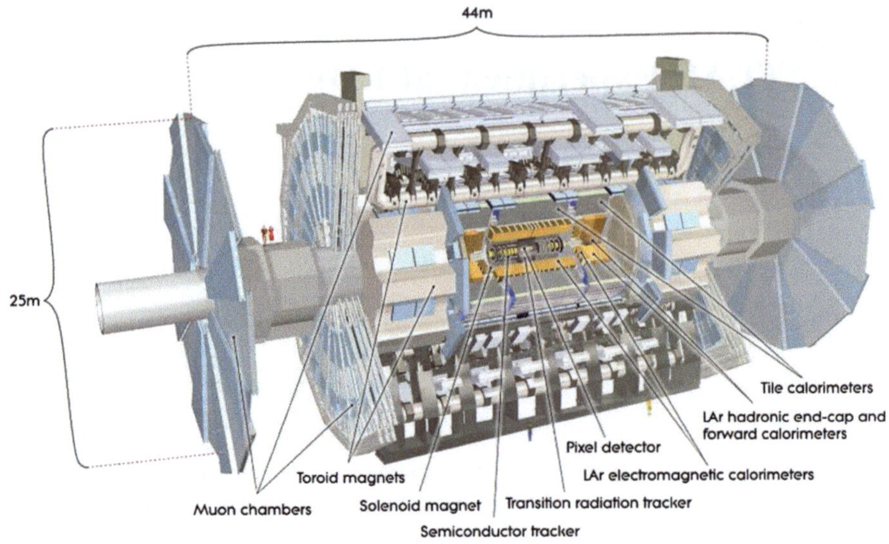

Fig. 2.1 Schematic view of the ATLAS detector

In the following, we will use a Cartesian right-handed coordinate system, with the origin in the nominal point of interaction, x axis pointing to the center of the LHC ring, and the y axis pointing upward. To define the direction of particles the following quantities are considered: the azimuthal angle $\phi = \arctan(y/x)$, and the pseudo-rapidity $\eta = -\ln(\tan(\theta/2))$, where θ is the polar angle defined from the z axis.

ATLAS is composed of different sub-detectors that we can divide in three different groups as treated in the following sections: the Inner Detector, the Calorimeters, and the Muon Spectrometer. A solenoidal magnetid field is used to bend the trajectories of the particles passing in the Inner Detector, while a toroidal magnetic field is used for the Muon Spectrometer. A sketch of the detector is presented in Fig. 2.1.

2.2.1 Inner Detector

The Inner Detector (ID) is the most inner part of ATLAS and occupies a cilindical volume around the LHC beam pipe. The ID has a radius of about 1.1 m and a length of 6.2 m along the beam pipe (see Fig. 2.2).

The ID has been designed to measure the properties of the charged particles generated in the collisions. Due to the high density of particles, a high granularity, a fast response and a good radiation resistence are needed to succesfully operated in the LHC environment. The ID is composed of three cylindrical concentric sub-detectors:

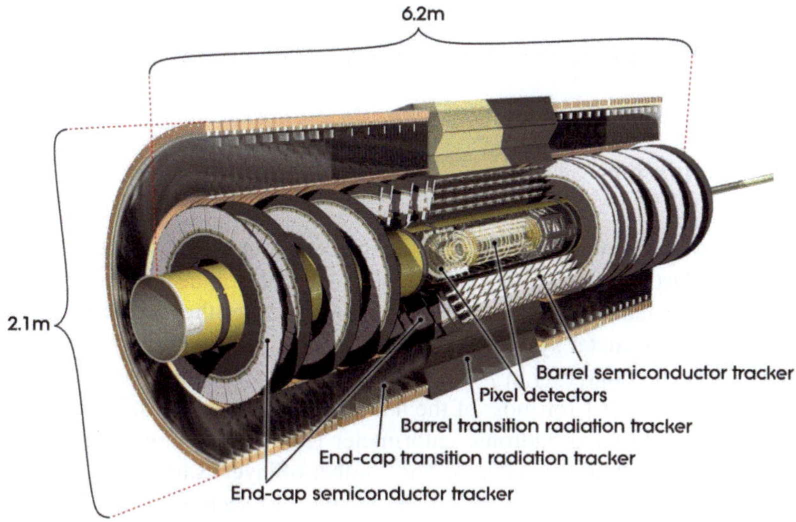

Fig. 2.2 Schematical view of the inner detector

- The Pixel detector measures charged particles using silicon sensors (pixels) and is the most inner part of the ID. The pixel sensors have a minimum size of $50 \times 400\ \mu m^2$, and provide a resolution of $10\,\mu m$ in the R-ϕ plane. Due to its high granularity, the Pixel detector has 80.4 milions read-out channels. This part of the ID mainly contributes to the precision vertex reconstruction.
- The Semiconductor Tracker (SCT) is a silicon microstrip detector, and is the middle part of the ID. This subdetector is composed of layers of stereo strips, so that approximately eight strips are crossed by each track and, since the position is determined from hits in overlapping strips, four space-points per track are usually available. Its sensors provide a resolution of $17\,\mu m$ in the R-ϕ plane, and make use of 6.2 milions read-out channels. This sub-detector contributes mainly to the momentum reconstruction.
- The Transition Radiation Tracker (TRT) consists of 4 mm diameter gaseous straw tubes interleaved with transition radiation material, and enables tracking for $|\ \eta\ | < 2.0$. It only provides R-ϕ information, for which it has an accuracy of $130\,\mu m$ per straw. The TRT is the most outern part of the ID and it has 351.000 readout channels. The TRT ease the track pattern recognition with its very large number of close hits (about 35 per track) and contributes to electron identification.

The first two are segmented in both r-ϕ and z, while the TRT is segmented only in r-ϕ. Using the combined information from the three subdetectors the transverse momentum resolution measured with cosmic muons [6] is:

$$\frac{\sigma_{p_T}}{p_T} = P_1 \oplus P_2 \times p_T, \tag{2.1}$$

where $P_1 = 1.6 \pm 0.1\,\%$ and $P_2 = (53 \pm 2)10^{-5}\,\text{GeV}^{-1}$. This translates in a resolution of 1.6 % for tracks with $p_T \sim 1\,\text{GeV}$ and of about 50 % for $p_T \sim 1\,\text{TeV}$.

2.2.2 Calorimeters

The ATLAS Calorimeter surrounds the ID and covers up to $|\eta| < 4.9$. It makes use of a lead and plastic tiles sampling calorimeter (TileCal) for the hadronic central part (up to $|\eta| < 1.7$) and a liquid Argon (LAr) calorimeter for the rest (see Fig. 2.3).

In total the calorimeter systems has 187,648 cells, and roughly 375,000 read-out channels. The cell granularity in ϕ and η varies between 0.025 in the electromagnetic (EM) central part, to 0.1 for most of the hadronic (HAD) sections, up to 0.4 for the most forward part of the hadronic calorimeter ($3.2 < |\eta| < 4.9$). The cells are organized in different layers in depth. The number of layers varies between three for the EM and the HAD central parts, to one for the EM section of the most forward part.

In the LAr calorimeter the liquid argon is the active medium, while the absorbers are either lead (for $|\eta| < 1.4$), or tungsten (for $1.4 < |\eta| < 4.9$ in the hadronic part), or copper for the rest. The liquid argon is kept at a temperature of 88 K with a cryogenic system, whose major components are housed between the LAr and the Tile calorimeters. Figure 2.4 (left) shows a sketch of the structure of one slice of the EM LAr calorimeter. Charged particles passing in the active material create couples of ions and electrons that drift in opposite directions by the presence of an electric

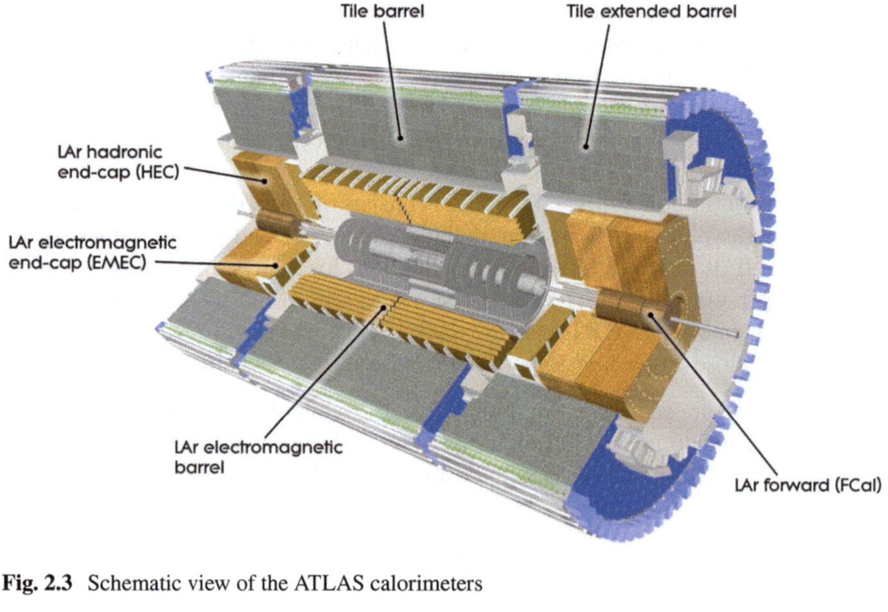

Fig. 2.3 Schematic view of the ATLAS calorimeters

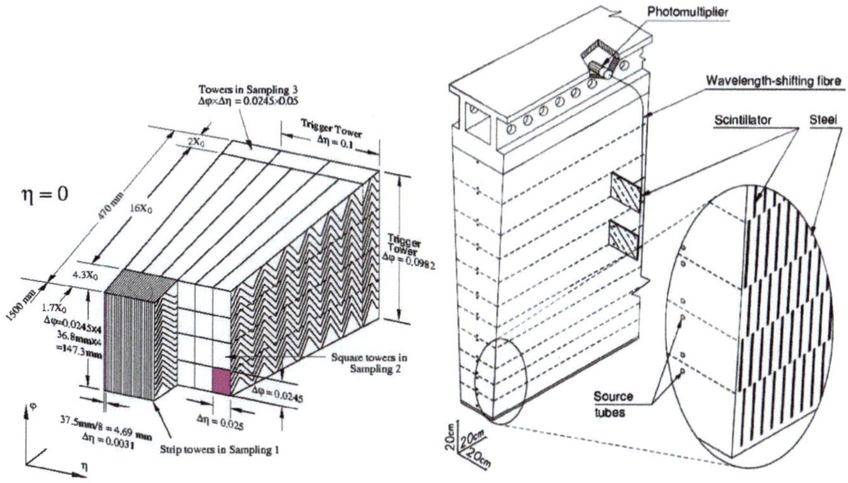

Fig. 2.4 Sketch of the structure of a LAr module (*left*) and a TileCal one (*right*)

field, and are collected by kapton electrodes. Different shapes of the kapton electrode surface have been used in order to provide a continuous calorimeter coverage in ϕ. For this reason an "accordion" geometry has been used for the central EM section (see Fig. 2.4 left), flat radial plates in the region $1.4 < |\eta| < 3.2$ both for the EM and HAD calorimeter, and tubes along the direction of the beam pipe for the most forward part. The relative energy resolution of the EM LAr calorimeter [7] is parametrized by:

$$\frac{\sigma_E}{E} = \frac{10\,\%}{\sqrt{E}} \oplus \frac{170\,\text{MeV}}{E} \oplus 0.7\,\% \tag{2.2}$$

TileCal is a sampling calorimeter with lead as absorbed and tiles of plastic scintillator as active material (see Fig. 2.4 right). The tiles emit light when charged particles pass through them. The light is collected by optical fibers and converted in pulses by photomultipliers. For TileCal the energy resolution is

$$\frac{\sigma_E}{E} = \frac{50\,\%}{\sqrt{E}} \oplus 3\,\%, \tag{2.3}$$

measured with test-beam data [8].

For both calorimeters the conversion factors between the reconstructed electronic signals and the deposited energy (called electromagnetic scale, or EM scale) have been determined by test-beam measurements with electrons in a range of 10–350 GeV. The electromagnetic scales have been validated with test-beam muons, cosmic muons, and E/p studies at collisions. Using collision data, the EM scale of the electromagnetic calorimeter has been corrected to reproduce the Z mass peak central value. The corrections are of the order of 1 %.

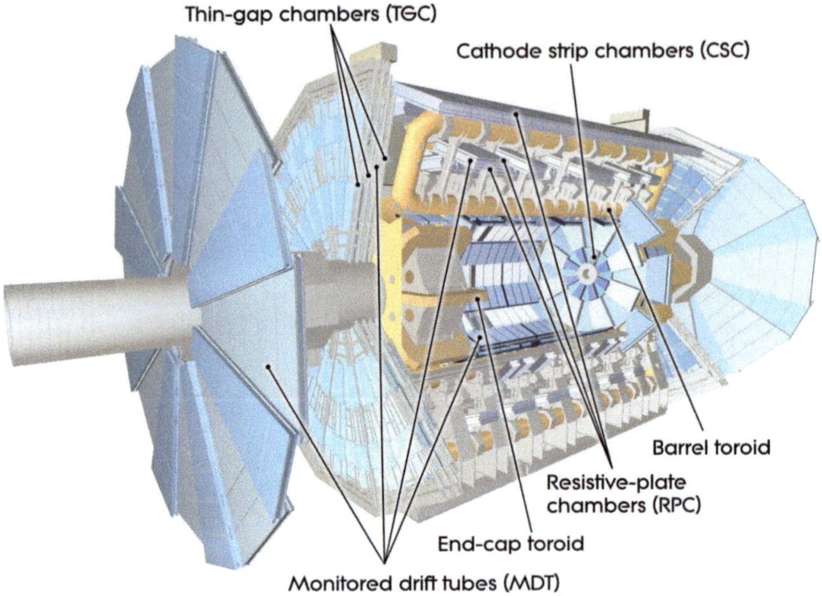

Fig. 2.5 Schematic view of the ATLAS muon spectrometer

The main aspects of the TileCal commissioning with cosmic muons, single beam data and first collision events are discussed in Appendix A.

2.2.3 Muon Spectrometer

The Muon Spectrometer is the most outer part of the the ATLAS detector and has been designed to identify and measure high momentum muons (see Fig. 2.5). The Muon Spectrometer is composed of four sub-detectors that make use of different technologies: Monitored Drift Tubes (MDT), Cathode Strip Chambers (CSC), Resistive Plate Chambers (RPC) and Thin Gap Chambers (TGC).

These sub-detectors are immersed in a magnetic field, generated by three toroids: one covers the central pseudo-rapidity range (approximately $|\eta| < 1.5$) providing a 0.5 T field, and other two, placed at higher pseudo-rapidity ($|\eta| > 1.5$), generating a 1 Tesla field. Each of them consists of eight coils assembled radially and symmetrically around the beam axis. The magnet configuration provides a field which bends the muon trajectories along the θ angle.

The MDT chambers perform a precision coordinate measurement in the bending direction of the air-core toroidal magnet, and therefore provide the muon momentum measurement for most of the eta range ($|\eta| < 2.7$). The basic detection element is a cylindrical aluminium drift tube of 3 cm diameter and a central wire at a potential of 3080 V. The drift tube is filled with a gas composed of Ar (93 %) and CO_2 (7 %). Muons passing in the tubes produce ionization charges that are collected on the wire. The reconstructed drift time provides a precise measurement of the minimum distance between the muon and the wire (with a typical resolution of 80 μm), and it's used to reconstruct the muon trajectory.

The CSC are multi-wire proportional chambers with cathodes segmented into strips, and are used at large pseudo-rapidities ($2.0 < |\eta| < 2.7$) in order to cope with the higher muon rate and background conditions. RPC and TGC are used for the trigger (see below) and measure also the muon ϕ coordinate.

2.2.4 Trigger System

In the last years, the LHC has been operating with a minimum time spacing of 50 ns between two bunch crossings. This translates into a maximum input rate of events of 20 MHz. The ATLAS trigger system is designed to bring this rate to 100 Hz, in order to record and store permanently the event information. The trigger system is organized in three levels: level 1 (LVL1), level 2 (LVL2), and Event Filter (EF). Each level refines the decision made at the previous step and, if necessary, applies additional selection criteria.

The LVL1 selects events with large missing energy or for high momentum muons, electrons, photons, tau and jets. The time available for the decision is 2.5 μs at most, therefore a simplified reconstruction of the physics objects is implemented. The LVL1 brings the rate of events to approximately 75 kHz.

The LVL2 has access to nearly all sub-detector information in a specific $\eta \times \phi$ region (called Region of Interest) around the objects selected at LVL1. The average event processing is 40 ms. The LVL2 brings the rate of events down to approximately 1 kHz.

The EF uses refined reconstruction algorithms that are very close to those used offline. The available time for event processing is 4 s in average, and the rate of events is reduced to approximately 100 Hz.

2.2.5 Luminosity Measurement

A precise luminosity determination is an essential ingredient for all physics analysis. In ATLAS, the luminosity is determined with several sub-detectors, each of them using different techniques [9, 10].

The luminosity can be expressed as:

$$L = \frac{n_b f_r n_1 n_2}{2\pi \Sigma_x \Sigma_y}, \tag{2.4}$$

where n_b the number of proton bunches crossing at the interaction point, and f_r the frequency of the LHC machine, n_1 and n_2 are the number of protons in the two LHC beams, and Σ_x and Σ_y characterize the horizontal and vertical profiles of the beams. In order to measure directly Σ_x and Σ_y, the two LHC beams are moved in steps of known distance, on both the horizontal and vertical direction. This special operation is called "van der Meer scan", or vdM scan. The absolute luminosity can therefore be measured at a vdM scan using Eq. 2.4, and knowing the currents of both LHC beams.

In order to relate the luminosity and μ, the average number of interaction per bunch crossing, the luminosity can be written as:

$$L = \frac{R_{inel}}{\sigma_{inel}} = \frac{\mu n_b f_r}{\sigma_{inel}}, \tag{2.5}$$

where σ_{inel} is the total inelastic cross section, R_{inel} is the rate of inelastic collisions, that can be written as product of n_b, f_r, and μ.

The visible number of interactions is defined as $\mu_{vis} = \epsilon\mu$, where ϵ is the efficiency of a particular sub-detector and a given algorithm to determine μ. Equation 2.5 can be rewritten as:

$$L = \frac{\mu_{vis} n_b f_r}{\sigma_{vis}}, \tag{2.6}$$

where σ_{vis} is the visible inelastic cross section, that can be retrieved at the vdM scan combining Eqs. 2.4 and 2.6. Therefore the luminosity is determined for a given measure of μ_{vis} through Eq. 2.6.

In order to measure μ_{vis} with a sub-detector, ATLAS primarily uses event counting algorithms, for which the number of events that satisfy a given criteria (typically a number of hits above a certain threshold) is compared with the total number of bunch crossings.

The default luminosity measurement in ATLAS for the 2011 and 2012 datasets was made using the Beam Conditions Monitor (BCM) [11]. The BCM is a fast device primarily designed to monitor background levels and issue a beam-abort request in case beam losses start to risk damage to ATLAS detectors. BCM consists of four small diamond sensors on each side of the ATLAS interaction point arranged around the beampipe in a cross pattern. In order to measure μ_{vis} with BCM, different event counting algorithms are used. The simplest algorithm requires a hit in either one of the two BCM arms (either at positive or negative η), and is called BCM_OR. At the vdM scan $\mu_{vis} \ll 1$, so that the probability to select an event with this algorithm is

$$\frac{N^{OR}}{N^{TOT}} = 1 - e^{-\mu_{vis}^{OR}}, \tag{2.7}$$

where N^{OR} is the number of bunch crossings that fulfilled the BCM_OR requirement, and N^{TOT} is the total number of bunch crossings.

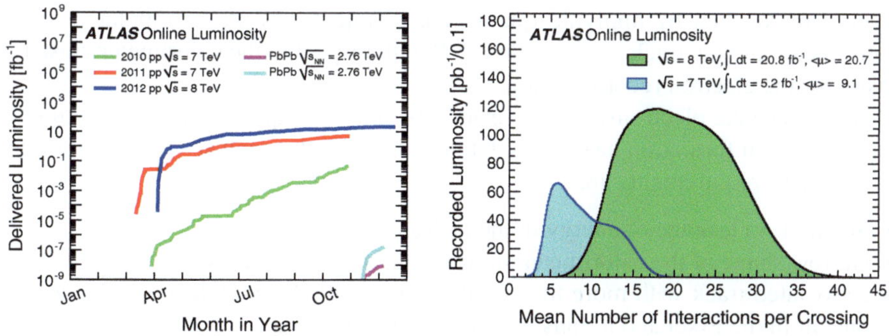

Fig. 2.6 Total integrated luminosity delivered by LHC in both p-p and Pb-Pb collisions from 2010 to 2012 (*left*). Mean number of interactions per bunch crossing in 2011 and 2012 p-p data (*right*)

For the p-p collisions realized until now the maximum instantaneous luminosity has been $7.7 \times 10^{33}\,\text{cm}^{-2}\,\text{s}^{-1}$, while for Pb-Pb and Pb-p collisions it has been 5.1×10^{26} and $1.1 \times 10^{29}\,\text{cm}^{-2}\,\text{s}^{-1}$ respectively. Figure 2.6 (left) shows the integrated luminosity of the various kinds of collision delivered from 2010 to 2012. Figure 2.6 (right) shows instead the μ distribution in the full 7 and 8 TeV datasets. Comparisons between the luminosity measured with BCM, TileCal and other sub-detectors are presented in Appendix B, which gives more details about the luminosity measurement using TileCal for both the 7 and 8 TeV data.

2.3 Reconstruction of Physics Object

This section describes the reconstruction of electrons, muons, jets, and missing transverse energy, giving emphasis to the aspects that are more important for the analyses presented in the next chapters.

2.3.1 Electrons

In the pseudo-rapidity region where the Inner Detector is operational ($|\eta| < 2.5$), electron candidates are defined by a matching between a track and a energy deposit in the EM calorimeter. The energy deposit is defined by a cluster of adjacent towers in $\eta \times \phi$. Each tower sums the energy in each cell in a region $\Delta\eta \times \Delta\phi = 0.025 \times 0.025$. The cluster of towers have size 3×7 in $\eta \times \phi$ for the calorimeter region $|\eta| < 1.4$, and 5×5 for the rest.

Tracks used for the electron candidates are reconstructed with an inside-out strategy. The track candidates built with the pixel detector and the first SCT layer are extrapolated to the outer part of the Inner Detector. Trajectories are fitted and fake tracks are rejected by applying quality cuts. Trajectories are extrapolated to the calorimeter and the track with smallest ΔR with respect to the calorimeter cluster is taken for the electron reconstruction.

To reject fake electrons, mainly from jets, quality requirements on calorimeter and tracking information have been defined [12]. For the analyses presented in this thesis, these requirements are:

- Small or no leakage of energy in the hadronic calorimeter,
- Narrow shape of the calorimeter shower consistent with real electrons,
- Associated track with more than 6 hits in the pixel and SCT layers, with at least one hit in the Pixel and a transverse impact parameter $d_0 < 5$ mm,
- Matching between track and calorimeter cluster with $\Delta\eta < 0.01$.

For the first two requirements the actual cuts depend on the electron energy and pseudo-rapidity [12].

The η, ϕ directions of the electron are taken from the associated track. For the 7 TeV analyses, the electron energy is taken from a weighted average between the track momentum and the cluster energy, while for 8 TeV analyses only the calorimeter information is used. The energy is corrected for dead material and for lateral and longitudinal leakage, with MC-based factors. These corrections have been validated in test-beams.

2.3.2 Muons

Muon candidates are reconstructed using Inner Detector and the Muon Spectrometer information [13]. The hits in each station of the Muon Spectrometer are combined to build track segments up to $|\eta| < 2.7$. For the analyses described in this thesis, muon candidates are reconstructed with two alternative procedures:

- "Combined" muon reconstruction. First, muon candidates are reconstructed using only Muon Spectrometer segments. The muon momentum measured in the Muon Spectrometer is corrected for the parametrized energy loss of the muon in the calorimeter, to obtain the muon momentum at the interaction point. Second, a match with a Inner Detector track is required and the momentum of the stand-alone muon is combined with the momentum measured in the Inner Detector.
- "Segment Tagged" muon reconstruction. A track in the inner detector is identified as a muon if the trajectory extrapolated to the Muon Spectrometer can be associated with straight track segments.

Combined muons are the highest purity muon candidates. Tagged muons give additional efficiency as they can recover muons which did not cross enough precision chambers to allow an independent momentum measurement in the Muon Spectrometer. Typical cases are low p_T muons ($p_T < 20$ GeV) that only reach the inner layer of precision chambers or less instrumented detector regions. In order to reject fake muons, all associated Inner Detector tracks are required to pass quality criteria based on the number of hits in the pixel, SCT and TRT systems.

2.3.3 Jets

In the analyses presented in this thesis, jets are reconstructed using the anti-kt algorithm [14], with jet radius parameter R = 0.4. The constituents of the jet finding algorithm are calorimeter clusters of energy, also called "topo-clusters"[15], defined as follow.

Jet constituents Topo-clusters are built out of neighboring calorimeter cells with significant energy deposit over the noise. The noise is measured for each cell independently, and it is defined as the expected RMS of the electronic noise for the current gain and conditions plus the contribution from pileup added in quadrature. In order to make topo-clusters, all cells with a signal to noise ratio |S/N| ≥ 4 are taken as seed cells for a topo-cluster formation. These cells are considered in descending order of S/N, and all neighboring cells with |S/N| ≥ 2 are added to the topo-cluster. Subsequently all cells adjacent to a topo-cluster cell are added, independently of their S/N. Final topo-clusters are treated as massless and their energy, at the electromagnetic scale, is the sum of the energies of the cells belonging to the topo-cluster.

Jet finding algorithm The anti-kt algorithm is a sequential recombination algorithm and is used in ATLAS in order to define the jets. For all constituents the algorithm computes the following quantities:

$$d_{ij} = \min(\frac{1}{k_{ti}^2}, \frac{1}{k_{tj}^2})\frac{\Delta R_{ij}^2}{R^2} \quad \text{and} \quad d_{iB} = \frac{1}{k_{ti}^2}, \tag{2.8}$$

where k_{ti} is the transverse momentum of constituent i, $R_{ij} = \sqrt{\Delta\eta^2 + \Delta\phi^2}$ between constituents i and j, and R a parameter of the algorithm that approximately controls the size of the jet. The distance d_{iB} is introduced in order to separate constituents coming from the hard interaction and those coming from proton remnants. The smallest distance is found, and if it is d_{ij}, constituents i and j are combined into one single object. If instead it is d_{iB}, constituent i is considered a jet and is removed from the list. The distances are recalculated with the remaining objects, and the process repeated until no constituent is left in the list. Once the process is finished, the jet four-momenta are defined by the vectorial sum of all the four-momenta of the topo-clusters belonging to it, so that the jet can have mass $m_{jet} > 0$. After this procedure, jets are defined with a minimum transverse momentum threshold p_{T}^{jet}, that is used as a scale to separate soft and hard physics.

Sequential recombination algorithms are very convenient because they are collinear and infrared safe to all orders in pQCD, as well as computationally fast. The anti-kt algorithm has been chosen because is particularly performant against pile-up, since it starts summing up constituents with higher momentum and produces jets with a conical structure.

Pile-up corrections Subsequently, corrections can be applied to subtract the energy contributions from multiple interactions from the same or previous bunch-crossings [16]. These corrections depend on the jet pseudo-rapidity and are parametrized as a function of the number of reconstructed vertex in the event, and on the average number of interactions μ.[1] The effect of pile-up on a central anti-kt jet with R = 0.4, is an over-estimation of the jet energy of about 0.2 GeV for each additional interaction in the same bunch-crossing. Multiple interactions in previous bunch-crossings, instead cause an under-estimation of the energy due to the particular LAr signal pulse shape. These two effects, therefore, tend to compensate. For the analysis of the 7 TeV data presented in this thesis, the effect of the pile-up corrections is very small and considered negligible; therefore these corrections are not applied. On the contrary, due to the higher pile-up level, corrections are applied in the analysis of the 8 TeV dataset.

Jet calibration The EM scale needs to be further calibrated to account for calorimeter non-compensation (the energy response to hadrons is lower than the response to electrons of the same energy), dead material (inactive regions of the detector where energy is lost) and leakage (energy deposits from particles which shower is not fully contained in the calorimeter). Moreover, corrections are needed for low momentum particles that are deflected by the magnetic field, and for energy losses in topo-cluster formation, and jet reconstruction. In order to correct for all these effects, a jet energy scale (JES) correction as a function of the jet energy and pseudo-rapidity is applied to jets at the EM scale [17]. The corrections are derived from di-jet MC samples produced with PYTHIA 6.423 with the AMBT1 tune. Calorimeter EM-scale jets are matched in ΔR with "truth" jets, that are reconstructed from stable particles in the final state,[2] excluding muons and neutrinos. Also the truth jets are built with the anti-kt algorithm with R = 0.4. The correction factors are based on the ratio of energies of the two matched jets E^{calo}/E^{truth}. It is important to mention that the calorimeter response in MC simulations has been extensively tested in both test beams and collision data. Figure 2.7 shows this ratio for different calibrated jet energies and as a function of the pseudo-rapidity of the reconstructed jet. The inverse of the response shown in each bin is equal to the average JES correction.

The reconstructed pseudo-rapidity direction can be biased due to the calorimeter response in different η regions. This effect is also corrected with MC-based correction factors. The η correction is about 0.01 for most of the calorimeter regions, and goes up to 0.05 in the transition regions between the different calorimeters. Jet four-momenta are also corrected to refer their kinematics to the primary vertex. This correction improves slightly ($\leq$1 %) the jet p_T response.

---

[1] The μ is measured every minute, so that it gives information about the average number of interactions in neighboring bunch-crossings.

[2] The final state in the MC generators is defined using all particles with lifetime above 10^{-11} s.

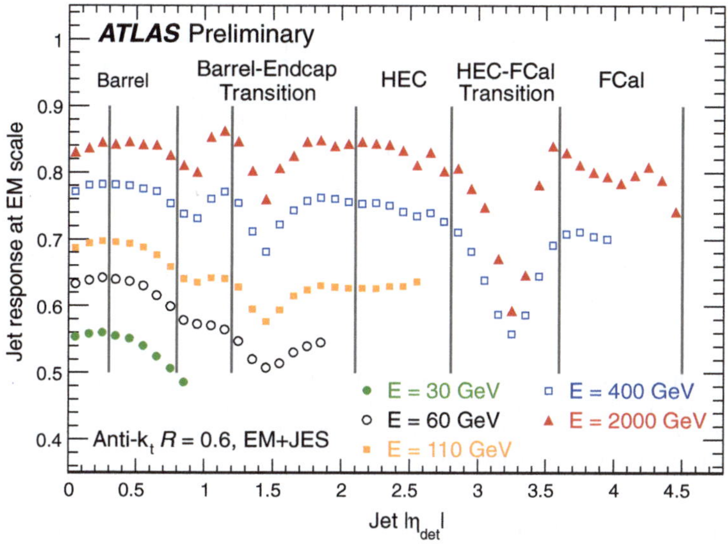

Fig. 2.7 Calorimeter response to jets before calibration in different pseudo-rapidity regions and for different energies. The inverse of the response shown in each bin is equal to the average jet energy scale correction

Jet energy scale uncertainty The jet energy scale (JES) uncertainty [18] comes mainly from the uncertainty on the single particle response, and has been estimated with E/p studies on isolated hadrons. Other leading uncertainties come from the limitations in the detector knowledge (such as the amount of dead material), and the physics models and parameters (mainly fragmentation and underlying event) in the MC event generator used to derive the JES corrections, the effects of having multiple interactions per bunch-crossing, and presence of close-by jets. Most of these uncertainties have been estimated for jets with $|\eta| < 0.8$ and then propagated to the more forward regions with the method of the p_T balance in di-jet events (see Ref. [19]). The reason is that the detector is better known in the central region, and that test-beam measurements to estimate the uncertainty due to the calorimeter response to single particles were only performed in the range $|\eta| < 0.8$.

For the 7 TeV data, the total JES uncertainty in absence of pile-up and near-by jets varies between 2.5 % for central high-p_T jets ($|\eta| < 0.8$, $60 < p_T < 800$ GeV) and 14 % for forward low-p_T jets ($3.2 < |\eta| < 4.5$, $20 < p_T < 30$ GeV). Close-by jets with $\Delta R < 0.7$ introduce an additional uncertainty of 2–3% dependently of the rapidity and the p_T of the jets. Multiple interactions in the same bunch-crossings introduce, for jets with $p_T = 30$ GeV ($p_T = 100$ GeV), an uncertainty of 0.5 % (0.1 %) for each reconstructed vertex. Figure 2.8 (left) shows the combination of all JES uncertainties for anti-kt jets with R = 0.6,[3] in $0.3 < |\eta| < 0.8$.

[3]Note that in this thesis jets are reconstructed with R = 0.4. JES uncertainties on jets with R = 0.4 and 0.6 are comparable.

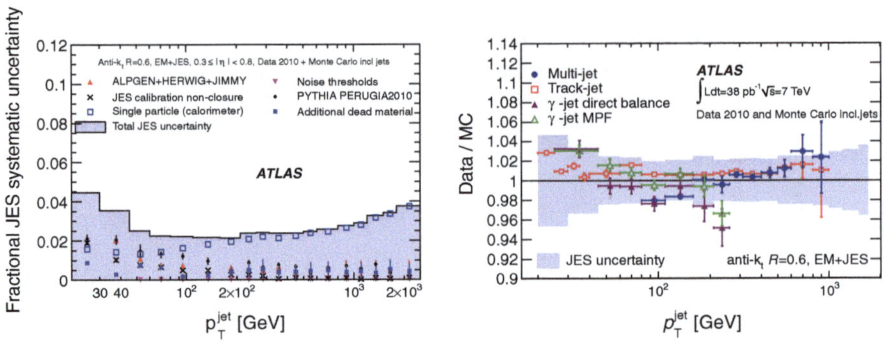

Fig. 2.8 The figure on the *left-hand side* shows the different contributions of the JES uncertainties for anti-kt jets with R = 0.6, in $0.3 < |\eta| < 0.8$, as a function of the jet p_T. The figure on the *right-hand side* shows the data-MC comparison on the jet response in various in-situ techniques, that is meant to validate the JES uncertainty

The JES is validated for jet p_T up to 1 TeV to the level of a few percent using several in situ techniques by comparing a well-known reference such as the recoiling photon p_T, the sum of the transverse momenta of tracks associated to the jet, or a system of low-p_T jets recoiling against a high-p_T jet. The JES systematic uncertainty determined from the combination of these in situ techniques are consistent with the one derived from single hadron response measurements over a wide kinematic range. Figure 2.8 (right) presents the ratio of data and MC jet responses in $|\eta| < 1.2$, compared with the JES uncertainty, showing compatible results.

Jet energy resolution The jet energy resolution (JER) [20] is measured with the in-situ techniques of the bisector [21] and the di-jet balance [22] methods, and the results are found in agreement within uncertainties. For jets with $p_T > 100$ GeV and $|\eta| < 0.8$ the JER is below 10 %, while is 15 % for jets with $p_T = 30$ GeV and $2.1 < |\eta| < 2.8$. Overall, the MC simulation of the JER agrees with the data within 10 %.

2.3.4 Missing Transverse Energy

In ATLAS there are different ways for calculating the missing transverse energy (E_T^{miss}). The most used method associates the calorimeter topo-clusters to reconstructed objects, such as jets, electrons, muons, taus, and photons, and calibrates them accordingly. In the next chapters we will use a E_T^{miss} made with a simpler calibration scheme, called "Local Cluster Weighting" (LCW) [23], that is more suitable for the analyses described in this thesis.

The LCW calibration method classifies the topo-clusters as either electromagnetic or hadronic, using the energy density and the shape of the cluster. Based on this classification energy corrections are derived from single pion MC simulations.

Dedicated corrections are derived for the effects of non-compensation, signal losses due to noise threshold effects, and energy lost in non-instrumented regions. They are applied to calorimeter clusters and are defined without reference to a given reconstructed object. They are therefore called local corrections. A special correction is made for the energy lost in the cryostat between the LAr electromagnetic calorimeter and the Tile calorimeter, which at a thickness of about half an interaction length can lead to signicant energy losses in hadronic showers.

After the topo-cluster four-momenta are calibrated the $E_{\mathrm{T}}^{\mathrm{miss}}$ is calculated form the vectorial sum of the p_{T} of all cells belonging to topo-clusters:

$$E_x^{\mathrm{miss}} = -\sum_{i=1}^{N_{cell}} E_i^{cell} \sin(\theta_i)\cos(\phi_i) \qquad E_y^{\mathrm{miss}} = -\sum_{i=1}^{N_{cell}} E_i^{cell} \sin(\theta_i)\sin(\phi_i) \tag{2.9}$$

The $E_{\mathrm{T}}^{\mathrm{miss}}$ performance can be studied by measuring the $E_{x,y}^{\mathrm{miss}}$ resolutions as function of the total transverse energy ΣE_T, which is reconstructed from the calorimeters as the scalar sum of the transverse energy of all cells. Figure 2.9 (left) shows the $E_{\mathrm{T}}^{\mathrm{miss}}$ resolution in minimum-bias events at 7 TeV. The $E_{\mathrm{T}}^{\mathrm{miss}}$ resolution is expressed as the root mean square of the $E_{x,y}^{\mathrm{miss}}$ distributions, and follows approximately a $k \times \sqrt{\Sigma E_T}$ function, with k being a parameter to be fitted. Three calibration schemes are compared in the figure: the LCW, the EM (topo-cluster four-momenta are used at the EM scale), and the "Global Cell Weighting", or GCW (see Ref. [24]). The latter applies cell-level weights that are based on the comparison between reconstructed and truth jets in MC. The figure shows how the LCW and GCW calibration schemes improve the $E_{\mathrm{T}}^{\mathrm{miss}}$ resolution. Figure 2.9 (right) compares instead the $E_{\mathrm{T}}^{\mathrm{miss}}$ resolution in data and MC events, showing a good agreement. For the simulation, samples of PYTHIA minimum-bias events have been used.

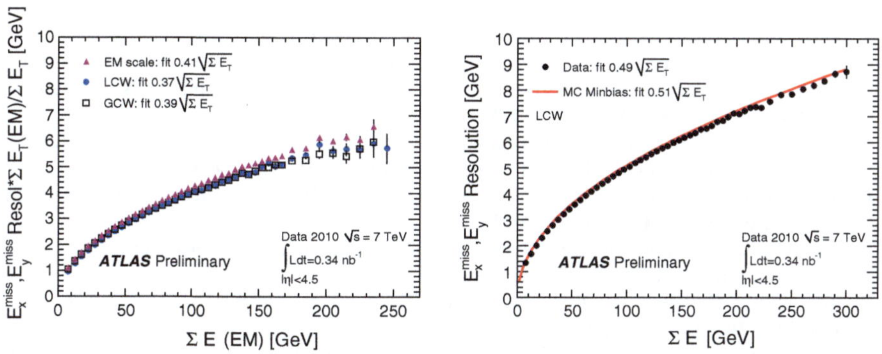

Fig. 2.9 The figure on the *left-hand side* shows the $E_{\mathrm{T}}^{\mathrm{miss}}$ resolution in minimum-bias events for the EM scale $E_{\mathrm{T}}^{\mathrm{miss}}$, and for LCW and GCW calibrated $E_{\mathrm{T}}^{\mathrm{miss}}$. Results are shown as a function of the EM scale ΣE_T. The figure on the *left-hand side* campares the resolution of the LCW $E_{\mathrm{T}}^{\mathrm{miss}}$ in data and MC events. Figures are taken from Ref. [24]

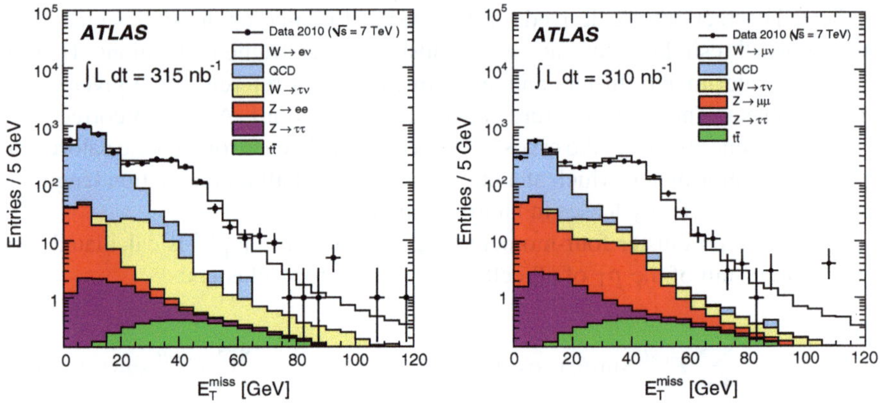

Fig. 2.10 The figure on the *left-hand side* shows E_T^{miss} distributions in an inclusive sample of events with one electron (*left*) or one muon (*right*) with $p_T > 20\,\text{GeV}$. The figures are taken from Ref. [25]

The LCW E_T^{miss} calibration scheme has been used successfully in W inclusive and W+jets cross section measurements [25, 26], providing a good understanding of these processes. Figure 2.10 (taken from reference [25]) shows the E_T^{miss} distributions in an inclusive sample of events with one electron (left) or one muon (right) with $p_T > 20\,\text{GeV}$. The figure shows how the distributions in data and MC are in reasonable agreement, both in the low E_T^{miss} region ($E_T^{miss} < 30\,\text{GeV}$) dominated by QCD multi-jet events, and for higher E_T^{miss} values dominated by W production.

References

1. P. Bryant, L. Evans (eds.), LHC machine. JINST **3**, S08001 (2008)
2. The ALICE Collaboration, The alice experiment at the CERN LHC. JINST **3**, S08002 (2008a)
3. G. Aad et al., The ATLAS experiment at the CERN large hadron collider. JINST **3**, S08003 (2008). doi:10.1088/1748-0221/3/08/S08003
4. The CMS Collaboration, The CMS experiment at the CERN LHC. JINST **3**, S08004 (2008b)
5. The LHCb Collaboration, The LCHb experiment at the CERN LHC. JINST **3**, S08005 (2008c)
6. The ATLAS Collaboration, Studies of the performance of the atlas detector using cosmic-ray muons. Eur. Phys. J. **C 71**, 1593 (2011)
7. The ATLAS Collaboration, Readiness of the ATLAS liquid argon calorimeter for LHC collisions. Eur. Phys. J. **C70**, 723–753 (2010a)
8. The ATLAS Collaboration, Readiness of the ATLAS tile calorimeter for LHC collisions. Eur. Phys. J. **C70**, 11931236 (2010b)
9. G. Aad et al., Luminosity determination in pp collisions at $\sqrt{s} = 7$ TeV using the ATLAS detector at the LHC. Eur. Phys. J. **C71**, 1630 (2011a). doi:10.1140/epjc/s10052-011-1630-5 (24 pages plus author list (36 pages total). 9 Figures, 10 Tables, submitted to Journal EPJC)
10. ATLAS Collaboration, Luminosity determination in pp collisions at $\sqrt{s} = 7$ TeV using the ATLAS detector in 2011 (ATLAS-CONF-2011-116), August 2011a
11. V. Cindro et al., The ATLAS beam conditions monitor. JINST **3**, P02004 (2008b)

12. G. Aad et al., Electron performance measurements with the ATLAS detector using the 2010 LHC proton-proton collision data. Eur. Phys. J. C **72**, 1909 (2012a) (Long author list—awaiting processing)
13. ATLAS Collaboration, Muon performance in minimum bias pp collision data at s = 7 TeV (ATLAS-CONF-2010-036), August 2011b
14. M. Cacciari, G.P. Salam, G. Soyez, The anti-k_t jet clustering algorithm. JHEP **04**, 063 (2008). doi:10.1088/1126-6708/2008/04/063
15. W. Lampl et al., Calorimeter clustering algorithms: description and performance (ATL-LARG-PUB-2008-002), April 2008
16. ATLAS Collaboration, Pile-up corrections for jets from proton-proton collisions at $\sqrt{s} = 7$ TeV in ATLAS in 2011 (ATLAS-CONF-2012-064), July 2012
17. G. Aad et al., Jet energy measurement with the ATLAS detector in proton-proton collisions at $\sqrt{s} = 7$ TeV (2011b)
18. The ATLAS Collaboration, Jet energy scale and its systematic uncertainty for jets produced in proton-proton collisions at $\sqrt{s} = 7$ TeV and measured with the atlas detector (Tech. Rep. ATL-CONF-2010-056), July 2010a
19. The ATLAS Collaboration, In-situ pseudorapidity intercalibration for evaluation of jet energy scale uncertainty using dijet events in proton-proton collisoins at $\sqrt{s} = 7$ TeV (ATL-CONF-2011-014), March 2011c
20. The ATLAS Collaboration, Jet energy resolution and selection efficiency relative to track jets from in-situ techniques with the atlas detector using proton-proton collisions at a center of mass energy $\sqrt{s} = 7$ TeV (Tech. Rep. ATL-CONF-2010-054), July 2010b
21. P. Bagnaia et al., UA2 Collaboration, Measurement of jet production properties at the CERN pp collider. Phys. Lett. B **144**, 283 (1984)
22. B. Abbott et al., D0 Collaboration, High-pT jets in pp collisions at $\sqrt{s} = 630$ GeV and 1800 GeV. Phys. Rev. **D64**, 032003 (2001)
23. T. Barillari et al. http://cdsweb.cern.ch/record/1112035ATL-LARG-PUB-2009-001
24. ATLAS Collaboration, Performance of the missing transverse energy reconstruction and calibration in proton-proton collisions at a center-of-mass energy of $\sqrt{s} = 7$ TeV with the ATLAS detector (ATLAS-CONF-2010-057) (2010c)
25. G. Aad et al., Measurement of the W $\rightarrow$ TeV with the ATLAS detector. JHEP **1012**, 060 (2010). doi:10.1007/JHEP12(2010)060 (38 pages plus author list (57 pages total), 16 figures, 15 tables)
26. The ATLAS Collaboration, Measurement of the production cross section for w-bosons in association with jets in pp collisions at $\sqrt{s} = 7$ TeV with the ATLAS detector. Phys. Lett. **B698**, 325–345 (2011)

Chapter 3
Search for New Phenomena in the Mono-jet Final State at $\sqrt{s} = 7\,\text{TeV}$

This chapter describes the mono-jet analysis performed with the full 2011 dataset of p-p collisions at $\sqrt{s} = 7\,\text{TeV}$. The analysis has been published in JHEP [1], and follows other two publications that made use of lower integrated luminosity: 33 pb^{-1} of data collected in 2010 [2], and the first 1 fb^{-1} of data collected in 2011 [3]. The analysis performed with $\sqrt{s} = 8\,\text{TeV}$ data collected in 2012 is the subject of the next chapter.

This chapter is organized as follows. Sections 3.1 and 3.2 present the various aspects of the ATLAS recorded dataset and simulated MC samples used for this analysis. The definition of the physics objects and the event selection criteria are detailed in Sects. 3.3 and 3.4, respectively. The background estimation is described in Sect. 3.5, and results are presented in Sect. 3.6. Section 3.7 discusses the interpretation of the results in the context of Large Extra Dimension models and WIMP pair production.

For the analysis of the full 2011 dataset, two alternative methods for Z/W + jet BG estimation have been developed within the ATLAS collaboration. The first method, presented in this chapter, follows closely the procedures used in the other ATLAS publications on searches in the mono-jet final state. The second method, detailed in Appendix E, makes use of slightly different proceedings, mainly regarding the selection of events to be used for control samples. The estimation of the BG processes and the level of systematic uncertainties are in good agreement between the two procedures.

Finally, the second method was adopted for the nominal results in the paper [1], while the method previously used was employed as a cross check.

3.1 Data Sample

This analysis makes use of the full 7 TeV dataset recorded by ATLAS in 2011. This corresponds to an integrated luminosity of 5.2 fb^{-1}. After applying basic data quality requirements on the data-taking conditions the remaining datasets consist

V. Rossetti, *Search for Exotic Mono-jet Events*, Springer Theses,
DOI 10.1007/978-3-319-22225-7_3

of 4.7 fb^{-1}. The maximum instantaneous luminosity increased from 1.3×10^{30} to $3.6 \times 10^{33}\,\text{cm}^{-2}\,\text{s}^{-1}$ along the year. This translated into an increase of the mean number of collisions per bunch crossing from 2.6 to 17.5.

Trigger selection Events are collected with the lowest unprescaled $E_{\text{T}}^{\text{miss}}$ trigger item, called "EF_xe60_verytight_noMu", which has the following thresholds at the three trigger levels:

- $E_{\text{T}}^{\text{miss}}(L1) > 50$ GeV
- $E_{\text{T}}^{\text{miss}}(L2) > 55$ GeV
- $E_{\text{T}}^{\text{miss}}(EF) > 60$ GeV

Details about the implementation of the $E_{\text{T}}^{\text{miss}}$ trigger can be found in Ref. [4]. The trigger algorithms use only calorimeter-based quantities with no corrections for identified muons.

LAr hole The detector conditions have also been changing during the year. The biggest detector problem that affected this analysis was the failure of 6 front-end boards (FEB) that correspond to adjacent areas of the barrel LAr calorimeter (see Sect. 2.2.2). This dead region of the calorimeter will be referred as the "LAr hole". Due to this problem it was not possible to reconstruct the energy deposited in the second and third layer of the calorimeter in the region with $-0.8 < \phi < -0.6$ and $0 < \eta < 1.4$. This problem concerned only a part of the dataset that corresponds to 1 fb^{-1} of integrated luminosity. During a LHC technical stop in July 2011, 4 out of the 6 non-functioning FEBs were recovered. After this fix only the energy in the third layer was unmeasured, resulting in a much smaller impact in the analysis. Dedicated cleaning requirements have been implemented in order to remove events affected by the LAr hole. This is described in the Sect. 3.4.

3.2 Monte Carlo Simulated Samples

MC samples of collision events are used in this analysis for the Standard Model (SM) background estimation and for the simulation of new physics processes.

For all the SM processes, the MC events are passed through the simulation of the detector made in GEANT4 [5]. Signal samples have been produced with ATLASFAST-II [6], that uses a simplified description of the interaction between particles and detector material. Single particle showers in the calorimeter are simulated using parametrized longitudinal and lateral energy profiles. For this reason, ATLASFAST-II is up to 100 times faster in terms of CPU time compared to the full GEANT4 simulation. Dedicated studies on jet and $E_{\text{T}}^{\text{miss}}$ shown that the ATLASFAST-II simulation reliably describes the detector response. Moreover, a comparison between fast and full simulation on the mono-jet signal samples shows no substancial differences, giving additional confidence to the use of ATLASFAST-II.

In order to account for the multiple interactions (pile-up) in the same and in the neighbor bunch crossing, all samples are generated with MC minimum bias interactions that are overlaid with the hard scattering event. The number of minimum bias interactions follows the distribution measured in the data for the analyzed dataset.

3.2.1 MC Generation of SM Processes

Z/W + jets production A set of Z/W + jets MC samples has been produced with ALPGEN [7] interfaced with HERWIG [8, 9] for the parton shower (PS), fragmentation and hadronization, and JIMMY [10] to simulate the underlying events. ALPGEN samples use the PDF set CTEQ6L1 [11] and the factorization and normalization scales are set to $\sqrt{M_{W/Z}^2 + p_{\mathrm{T}}^2}$, where p_{T} is the scalar sum of the p_{T} of the outgoing partons.

$W(\rightarrow \ell\nu)$ + jets and $Z(\rightarrow \nu\nu)$ + jets samples are generated with up to six outgoing partons with LO matrix elements, while the $Z/\gamma^*(\rightarrow \ell\ell)$ + jets samples are generated with up to five partons. The MLM technique (see Sect. 1.1) is used to match matrix elements to PS evolution.

Another set of Z/W + jets samples produced with SHERPA [12] has been used in the analysis to assess systematic uncertainties. The CT10 PDF set has been used for this generation. All Z/W + jets samples (from both ALPGEN and SHERPA) are initially normalized to the inclusive Drell-Yan and $W(\rightarrow \ell\nu)$ cross sections calculated at NNLO in perturbative QCD with the FEWZ program [13], using MSTW2008 PDFs.

Top production The production of single top and $t\bar{t}$ has been generated with MC@NLO [14]. A top mass of 172.5 GeV and the NLO PDF set CTEQ6.6 [15] have been used for the generation. MC@NLO is interfaced with HERWIG and JIMMY, for parton shower, hadronization, and underlying event.

Inclusive jet production Jet production from QCD-only interactions is simulated with PYTHIA [16]. It's worth to mention that the QCD multi-jet background is determined from data, and that these MC samples are only used to build distributions for the plots.

Di-bosons production Samples of WW, WZ and ZZ production generated with HERWIG and SHERPA have been used for the di-boson background estimation. NLO cross sections calculated using MCFM [17] have been employed for the normalization.

3.2.2 MC Generation of Graviton Production in ADD Scenario

MC samples of graviton production in association with a parton (see Sect. 1.4.1) are generated using PYTHIA, and the PDF set CTEQ6.6 has been considered for final results.[1] A low energy effective field theory has been implemented as in Ref. [19] considering the contributions of different graviton mass modes. The renormalization and factorization scale have been set to $\sqrt{M_G^2 + p_{\mathrm{T}}^2}$ where p_{T} is the transverse momentum of the outgoing parton. Samples have been generated with number of extra dimensions n between two and six, and the scale of the effective field theory M_D ranging between 2.5 and 3.5 TeV.[2] For the interpretation of the results in terms of graviton production, events are normalized both to the LO and NLO cross sections calculated by the authors of Ref. [20], for the phase space explored in this analysis.

3.2.3 MC Generation of WIMP Pair Production

The production of WIMP pairs plus jets has been simulated with MADGRAPH5 [21] with LO matrix elements for up to two outgoing partons from ISR. The PDF set used for these MC samples has been CTEQ6L1. The renormalization and factorization scales are set to the sum of $\sqrt{m^2 + p_{\mathrm{T}}^2}$ for all the particles in the final state. PYTHIA is used for PS and hadronization, and the MLM prescription is employed to match matrix elements to PS.

In this analysis, WIMPs are assumed to be Dirac fermions and we consider the interaction operators D1, D5, D8, D9 and D11, described in Sect. 1.4.2, following the convention in [22]. For each operator, the MC samples have been produced with WIMP masses ranging from 10 GeV to 1.3 TeV. The interactions between WIMPs and quarks are considered flavour-universal for the first two quark generations.

3.3 Object Definition

Details on the reconstruction and calibration of electrons, muons, jets, and $E_{\mathrm{T}}^{\mathrm{miss}}$ were given in Sect. 2.3.

Jet candidates are reconstructed from calorimeter topo-clusters using the anti-kt algorithm [23] with the distance parameter $R = 0.4$. The jet energy is corrected with p_{T} and η dependent calibration factors [24]. After the calibration, only the jets with

[1]The ADD MC samples are in fact produced with PDF set MRST2008 LO** [18]. Event yields as well as cross section and PDF systematic uncertainties are obtained by re-weighting the MC samples to the CTEQ6.6 set.

[2]The acceptance of this signal does not depend on M_D so that limits can be extrapolated to higher values of M_D.

$p_T > 30$ GeV and $|\eta| < 4.5$ are considered for this analysis.[3] Jets are ordered in p_T so that the leading jet is the jet with the highest p_T.

The E_T^{miss} is defined by the vectorial sum of all calorimeter topo-clusters with $|\eta| < 4.5$. Each cluster is calibrated with factors that take into account the different response of the calorimeters to hadrons compared to electrons or photons, as well as dead material and out-of-cluster energy losses [25, 26]. In this analysis the E_T^{miss} is only calorimeter-based and no correction for potential identified muons is applied.

Electron candidates are reconstructed from a calorimeter cluster of energy associated with a good quality track. Electrons are required to pass the *medium* [27] selection criteria, based on the electron shower shape and associated track. In this analysis, electrons are required to have $p_T > 20$ GeV and $|\eta| < 2.47$. In order to resolve ambiguities between electrons and jets, an overlap removal procedure has been applied. If a jet and an electron have $\Delta R(\text{jet, el.}) < 0.2$ the electron is kept in the analysis, and the jet is considered a fake and therefore removed from the jet list. Instead, if the electron-jet distance is $0.2 < \Delta R < 0.4$, the electron is considered part of the jet, and therefore it is removed from the electron list, while the jet is kept in the analysis.

Muon candidates are reconstructed from either associating a stand-alone muon spectrometer track to an inner detector track, or from an inner detector track that is confirmed by a directional segment in the muon spectrometer [28]. In this analysis, muons are selected with $p_T > 7$ GeV and $|\eta| < 2.5$. They are required to be isolated: the scalar p_T sum of tracks within $\Delta R < 0.2$ around the muon track must be less than 1.8 GeV.

MC events are corrected to have the same electron and muon reconstruction efficiency as in data. The correction factors, in bins of p_T and η, have been retrieved from inclusive $W(\rightarrow e\nu)$, $W(\rightarrow \mu\nu)$, $Z(\rightarrow ee)$ and $Z(\rightarrow \mu\mu)$ events. These factors typically differ from unity by less than 0.5 %.

3.4 Event Selection

Data and MC events are required to pass the following selection criteria:

- Events are required to have $E_T^{miss} > 120$ GeV and a leading jet with $p_T > 120$ GeV and $|\eta| < 2$.
- Events with more than two jets (with $p_T > 30$ GeV and $|\eta| < 4.5$) are rejected.
- In order to reduce the multi-jet background from QCD interactions, the second leading jet (if any) is required to have $\Delta\phi(\text{jet2}, E_T^{miss}) > 0.5$.
- Reconstruction of a primary vertex with at least two tracks of $p_T > 0.4$ GeV. This requirement rejects non-collision events.

[3] The only exception to this rule is for the rejection of non-collision events due to coherent noise or fake signal in the calorimeter, for which jets are defined with $p_T > 20$ GeV.

Table 3.1 Summary of the selection requirements for the four signal regions

Signal regions	SR1	SR2	SR3	SR4
Common requirements	Vertex + jet cleaning + LAr hole cleaning + $N_{jets} \leq 2$ + $\|\eta^{jet1}\| < 2$ + $\Delta\phi$(jet2, $E_{\mathrm{T}}^{\mathrm{miss}}$) > 0.5 + lepton veto			
$E_{\mathrm{T}}^{\mathrm{miss}}>$ (GeV)	120	220	350	500
$p_{\mathrm{T}}^{jet1} >$ (GeV)	120	220	350	500

- No jet with $p_{\mathrm{T}} > 20\,\mathrm{GeV}$[4] and $|\eta| < 4.5$ presenting anomalous behavior of the jet quantities, such as the timing, the shape of the electronic pulses used for the energy reconstruction, or the fraction of energy deposited in the different calorimeter layers. These requirements spot events with fake jets as well as coherent noise or electronic burst in the calorimeters (see Ref. [29]).
- In order to remove remaining events from beam-related backgrounds, the leading jet is required to have $f_{CH} = \Sigma p_{\mathrm{T}}^{track}/p_{\mathrm{T}}^{jet} > 0.02$, where $\Sigma p_{\mathrm{T}}^{track}$ is the scalar sum of the p_{T} of all tracks with $\Delta R < 0.4$ from the jet axis. Furthermore the leading jet is required to have $f_{EM} \geq 0.1$, where f_{EM} is the fraction of jet energy deposited in the electromagnetic calorimeter.
- As described in Sect. 3.1, part of the data has been affected by an electronic failure in the LAr calorimeter. Events are rejected if a jet is reconstructed in the affected area and if it points in the direction of the $E_{\mathrm{T}}^{\mathrm{miss}}$ ($\Delta R(jet, E_{\mathrm{T}}^{\mathrm{miss}}) < 0.4$). This requirement only rejects a few percent of the events in the affected part of the dataset and the impact on the final results is therefore negligible.
- Events with identified muons or electrons are rejected. This cut reduces the background from all processes with isolated leptons, mainly $W(\rightarrow \ell\nu)+$jet and $Z(\rightarrow \ell\ell)+$jet, but also top and di-boson production.

Four different signal regions (SRs) are defined, with cuts on $E_{\mathrm{T}}^{\mathrm{miss}}$ and leading jet p_{T} of 120, 220, 350 and 500 GeV. The selection requirements are summarized in Table 3.1.

Trigger efficiency For the selection requirements of SR1, the trigger item used to collect events (see Sect. 3.1) is not fully efficient. To determine the trigger efficiency we select an unbiased data sample collected with a different trigger (called "EF_mu18_medium") whose algorithms are based on Muon Spectrometer quantities. The total integrated luminosity of this sample is 4.5 fb^{-1}. Events are then required to pass the mono-jet selection of SR1, plus the identification of a muon with $p_{\mathrm{T}} > 20\,\mathrm{GeV}$.

Figure 3.1 shows the trigger efficiency as a function of the $E_{\mathrm{T}}^{\mathrm{miss}}$. The trigger item used for this analysis is found to be more than 98 % efficient for events in SR1, while it is fully efficient for the other SRs. For comparison a $W(\rightarrow \mu\nu)+$jet MC sample generated with ALPGEN has been used. MC events are required to pass the same

[4]Only for this requirement, jets are selected with $p_{\mathrm{T}} > 20\,\mathrm{GeV}$. Otherwise, jets are defined with the threshold $p_{\mathrm{T}} > 30\,\mathrm{GeV}$.

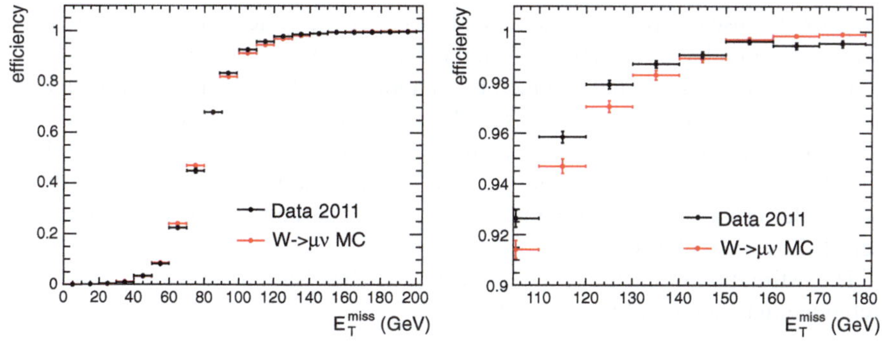

Fig. 3.1 Efficiency of the trigger used for this analysis (called "EF_xe60_verytight_noMu") as a function of the offline $E_{\mathrm{T}}^{\mathrm{miss}}$. An independent trigger selection ("EF_mu18_medium") has been used for this study. Data (*in black*) is compared to a $W(\rightarrow \mu\nu)$ Alpgen MC (*in red*). On the *right*, a *zoomed view* of the *left plot* is shown. Figures are taken from Ref. [30]

event selection of the data. Trigger efficiencies derived from data and from MC agree within 1 %. This difference is taken into account in the BG estimation and has a small impact in the BG determination.

3.5 Estimation of the Background Contributions

3.5.1 Z/W+jets Production

The production of a Z or W boson in association with jets constitutes the main background (BG) of this analysis (~97 % of the total BG). Controlling the total number of events and also the shape of the kinematic distributions (such as $E_{\mathrm{T}}^{\mathrm{miss}}$, p_{T} of the jets, etc.) is crucial for reaching a good sensitivity for mono-jet signals coming from new physics.

A pure MC prediction of these processes would suffer from large theoretical uncertainties (up to 40 %) originating, for example, from the choice of the PDFs and the renormalization and factorization scales, as well as from experimental uncertainties related to the determination of the absolute JES and luminosity, among others. These systematic uncertainties affect both the absolute normalization and the shape of the predicted distributions. Nevertheless, ATLAS results on boson+jet(s) production [31, 32] demonstrate that the nominal MC predictions describe the data reasonably well, and that the data has the potential to constrain the size of the systematic uncertainty on the MC.

This section describes a data-driven method developed in order to estimate the Z/W+jet contribution to the signal region (SR). The method has been designed to minimize the use of MC information, and to reduce the systematic uncertainty using as much as possible informations from the data in control regions (CRs).

The dominant BG is the irreducible component of the Z +jets in which $Z(\rightarrow \nu\nu)$ decays generate large $E_{\mathrm{T}}^{\mathrm{miss}}$. The W +jets BG is composed by $W(\rightarrow \tau\nu)$, $W(\rightarrow \mu\nu)$ and $W(\rightarrow e\nu)$ events in which no electrons or muons are identified. The $Z(\rightarrow \ell\ell)$ +jets contribution is much smaller compared to the others due to the smaller cross section and the presence of two charged leptons in the final state (rejected by the lepton veto).

The estimation of each of these contributions has been done with a data-driven technique that can be split in three steps:

- Define the CR to select Z/W +jet events in data. CRs are defined to have no overlap with the SR and to have none or negligible contributions from monojet-like signals.
- Using MC events, build the transfer factors (TF) (defined below), that converts observations in the CR into background estimates for the SR.
- Multiply the TF to the number of events observed in data in the CR, to get the data-driven estimation of the process in the SR.

In the following, we will first define the CRs and the TFs, and then we will detail the systematic uncertainties on the BG estimation.

3.5.1.1 Control Regions

In this analysis, four different control regions (CRs) have been considered: two inclusive CRs with at least one identified muon or one identified electron, and two specific CRs to select $W(\rightarrow \mu\nu)$ or $Z(\rightarrow \mu\mu)$ decays.

The inclusive electron CR is defined by inverting the electron veto and applying all the other selection cuts of the SR. This means that all the events with at least one electron will enter in the electron CR.

Events in the inclusive electron CR are mainly $W(\rightarrow e\nu)$ +jets, with a contamination from $W(\rightarrow \tau\nu)$ +jets with $\tau \rightarrow e\nu\nu$ (~20 %), top (~3 %), di-boson (<1 %) and $Z(\rightarrow \tau\tau)$ +jets (<1 %). It is worth to mention that $Z(\rightarrow ee)$ +jets events are not passing the selection because of the large $E_{\mathrm{T}}^{\mathrm{miss}}$ requirement.

Similarly, the inclusive muon CR is defined by inverting the muon veto and applying all other selection cuts of the SR. This CR is mainly composed by $W(\rightarrow \mu\nu)$ +jets with smaller fractions of $Z(\rightarrow \mu\mu)$ +jets (~10 %), $W(\rightarrow \tau\nu)$ +jets (~ 10 %), but also top (~5 %), $Z(\rightarrow \tau\tau)$ +jets (~3 %), and di-boson (~1 %).

The $W(\rightarrow \mu\nu)$ CR and the $Z(\rightarrow \mu\mu)$ CR are defined in order to select the two processes separately. The $W(\rightarrow \mu\nu)$ CR is defined selecting events with only one muon and applying an extra selection cut on the reconstructed W transverse mass $40 < M_T < 100$ GeV, defined as

$$M_T = \sqrt{2\, p_{\mathrm{T}}\, E_T^{\mathrm{miss},\mu}\, (1 - \cos\Delta\phi(\mathrm{muon}, E_T^{\mathrm{miss},\mu}))},$$

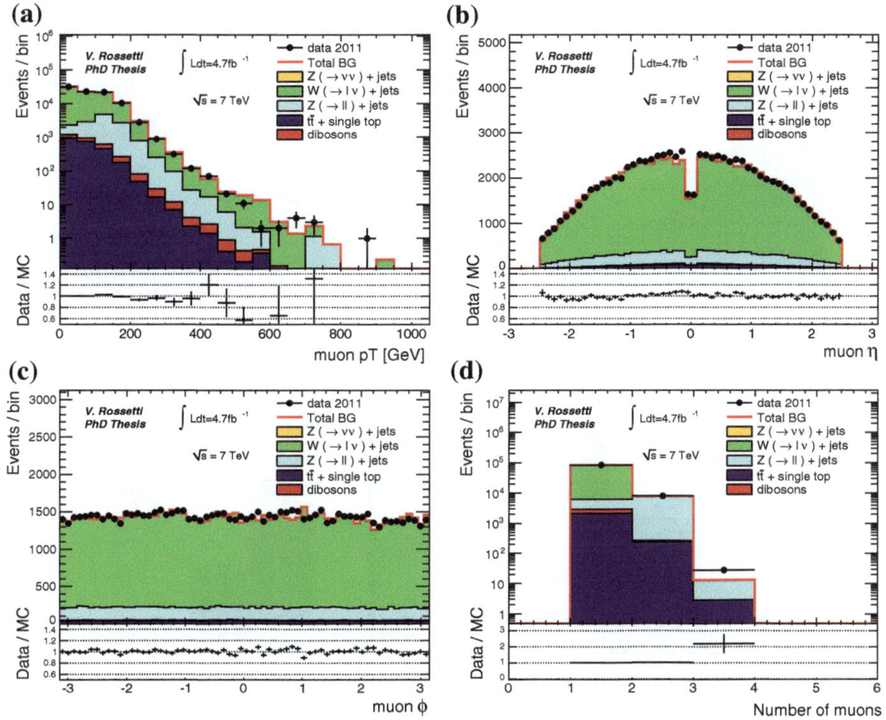

Fig. 3.2 Kinematic distributions of the identified muons in the inclusive muon CR for the selection cuts of region 1 ($E_{\mathrm{T}}^{\mathrm{miss}}$, jet1 $p_{\mathrm{T}} > 120\,\mathrm{GeV}$)

where $E_T^{\mathrm{miss},\mu}$ is the missing transverse energy (calorimeter based, as defined in Sect. 3.4) plus a term to consider the four-momentum of the reconstructed muon. In this way $E_T^{\mathrm{miss},\mu}$ is the best estimate of the transverse momentum of the neutrino escaping detection.[5] Similarly, the $Z(\rightarrow \mu\mu)$ CR is defined selecting events with exactly two muons with an invariant mass within $76 < M_{\mu\mu} < 116\,\mathrm{GeV}$. It has to be noticed that the $W(\rightarrow \mu\nu)$ CR and the $Z(\rightarrow \mu\mu)$ CR are in fact two sub-samples of the inclusive muon CR.

CRs and SRs are defined by the presence or the absence of electrons and muons, therefore a good knowledge of the identification of these object is crucial for a precise BG prediction. Figures 3.2 and 3.3 show the main kinematic distribution of the reconstructed leptons for SR1 ($E_{\mathrm{T}}^{\mathrm{miss}}$, jet1 $p_{\mathrm{T}} > 120\,\mathrm{GeV}$). Figure 3.4 shows the distributions of M_T and $M_{\mu\mu}$. The M_T distribution is shown for events with only one identified muon, and the $M_{\mu\mu}$ for those with exactly two muons.

For each SR, a corresponding set of four CRs (two inclusive and two for the $W(\rightarrow \mu\nu)$ and $Z(\rightarrow \mu\mu)$ selection) is defined with the same $E_{\mathrm{T}}^{\mathrm{miss}}$ and jet selection.

[5] Only in the definition of the transverse mass M_T, the $E_{\mathrm{T}}^{\mathrm{miss}}$ is defined with a correction term to account for identified muons. Otherwise, the $E_{\mathrm{T}}^{\mathrm{miss}}$ is only calorimeter based.

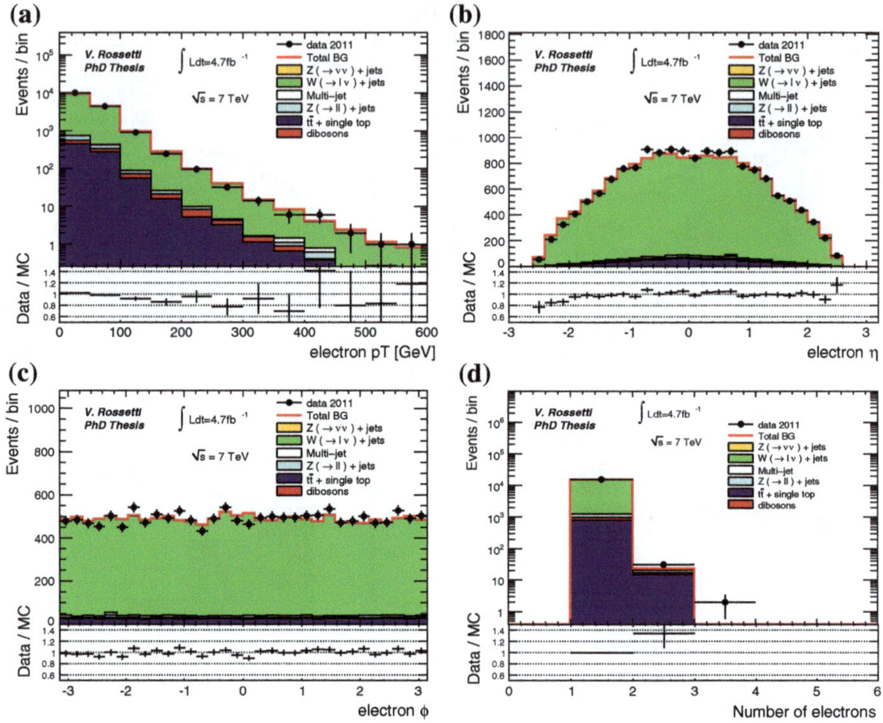

Fig. 3.3 Kinematic distributions of the identified electrons in the inclusive electron CR for the selection cuts of region 1 ($E_{\mathrm{T}}^{\mathrm{miss}}$, jet1 $p_{\mathrm{T}} > 120\,\mathrm{GeV}$)

The BG estimation for SR1 ($E_{\mathrm{T}}^{\mathrm{miss}}$, jet1 $p_{\mathrm{T}} > 120\,\mathrm{GeV}$) has been made from the four CRs with $E_{\mathrm{T}}^{\mathrm{miss}}$, jet1 $p_{\mathrm{T}} > 120\,\mathrm{GeV}$. Similarly for SR2, SR3 and SR4 the estimations make use of CRs with $E_{\mathrm{T}}^{\mathrm{miss}}$, jet1 $p_{\mathrm{T}} > 220$, 350 and 500 GeV respectively.

Figures 3.5 and 3.6 present the measured $E_{\mathrm{T}}^{\mathrm{miss}}$ and jet distributions in the inclusive electron and inclusive muon CRs, respectively, compared to the MC predictions. Similarly, Figs. 3.7 and 3.8 show the $E_{\mathrm{T}}^{\mathrm{miss}}$ and jet distributions as measured in the exclusive $W(\rightarrow \mu\nu)$ and $Z(\rightarrow \mu\mu)$ CRs. In all these plots, and only for illustration to put emphasis on the shape, the MC samples are normalized to the data including a global scale factor. These normalization factors for all the CRs are collected in Table 3.2. The factors have been calculated after the Z/W + jets MC samples have been normalized to the inclusive NNLO Drell-Yan and inclusive $W(\rightarrow \ell\nu)$ cross sections (see Sect. 3.2), and after the subtraction of the other processes (top and di-boson production). All the distributions show a reasonable agreement between data and MC.

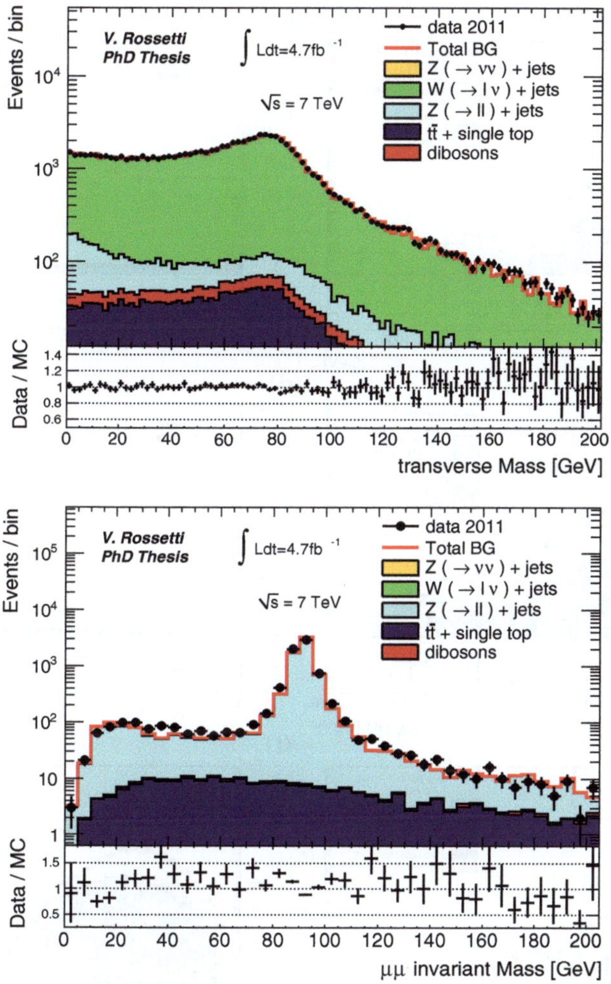

Fig. 3.4 The distributions of M_T and $M_{\mu\mu}$. The exclusive $W(\rightarrow \mu\nu)$ CR is obtained requiring $40\,\text{GeV} < M_T < 100\,\text{GeV}$. Instead the $Z(\rightarrow \mu\mu)$ exclusive CR is obtained requiring $76\,\text{GeV} < M_{\mu\mu} < 116\,\text{GeV}$

3.5.1.2 Transfer Factors Method

As already mentioned, the aim of the transfer factor method is to reduce as much as possible the uncertainties on the BG estimation exploiting the information of the data in the CRs. In order to estimate the BG contribution of a given process in the SRs the number of data events in a CR is corrected with MC-based factors (called transfer factors or simply TFs). As an example, in order to estimate the $Z(\rightarrow \nu\nu)+$jets contribution to the SR we can clearly make use of the events in the $Z(\rightarrow \mu\mu)+$jets CR. Since the $E_{\mathrm{T}}^{\mathrm{miss}}$ is calorimeter based, $Z(\rightarrow \nu\nu)+$jets events in the SR and

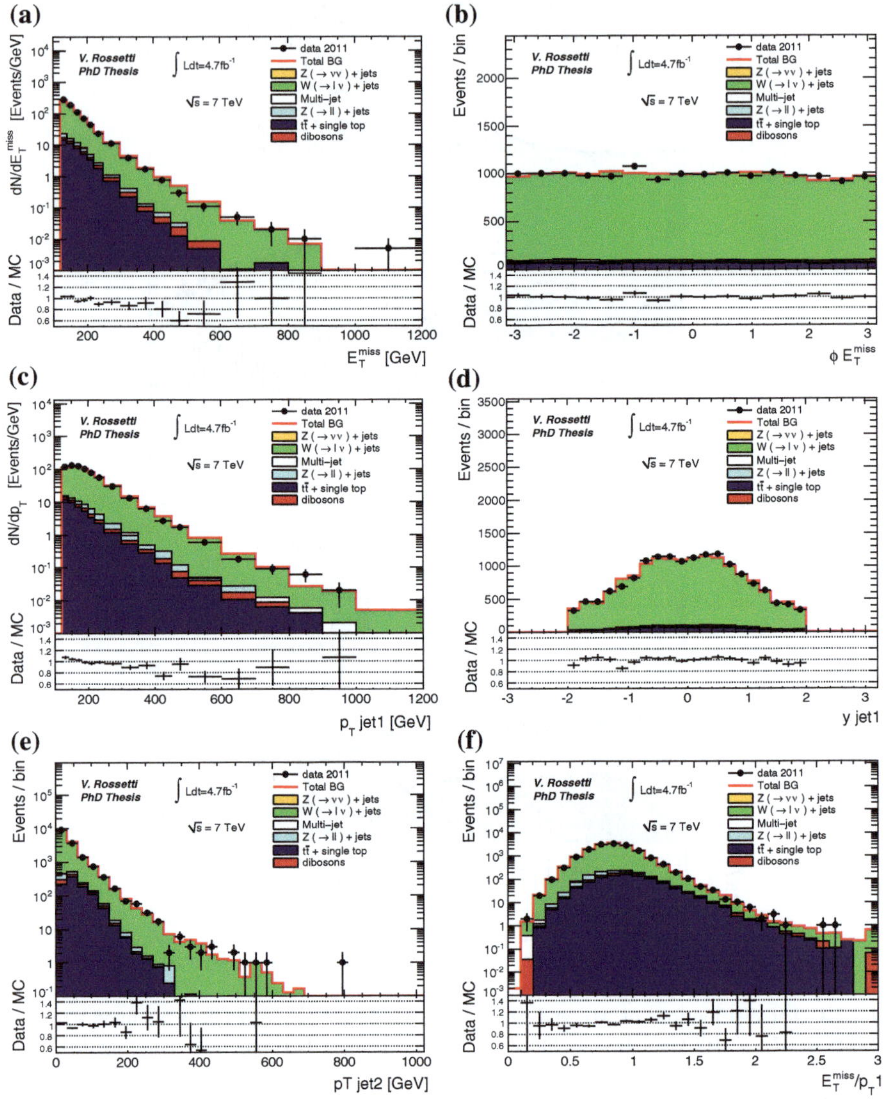

Fig. 3.5 Kinematic distributions of the events in the inclusive electron CR for the selection cuts of region 1 ($E_{\mathrm{T}}^{\mathrm{miss}}$, jet1 $p_{\mathrm{T}} > 120\,\mathrm{GeV}$)

$Z(\rightarrow \mu\mu) +$jets events in the CR have a very similar behavior.[6] The differences are the related to:

- the branching ratios of the Z decays,
- the muon identification,

[6]This is the reason why the $E_{\mathrm{T}}^{\mathrm{miss}}$ has been defined without any correction for the muons.

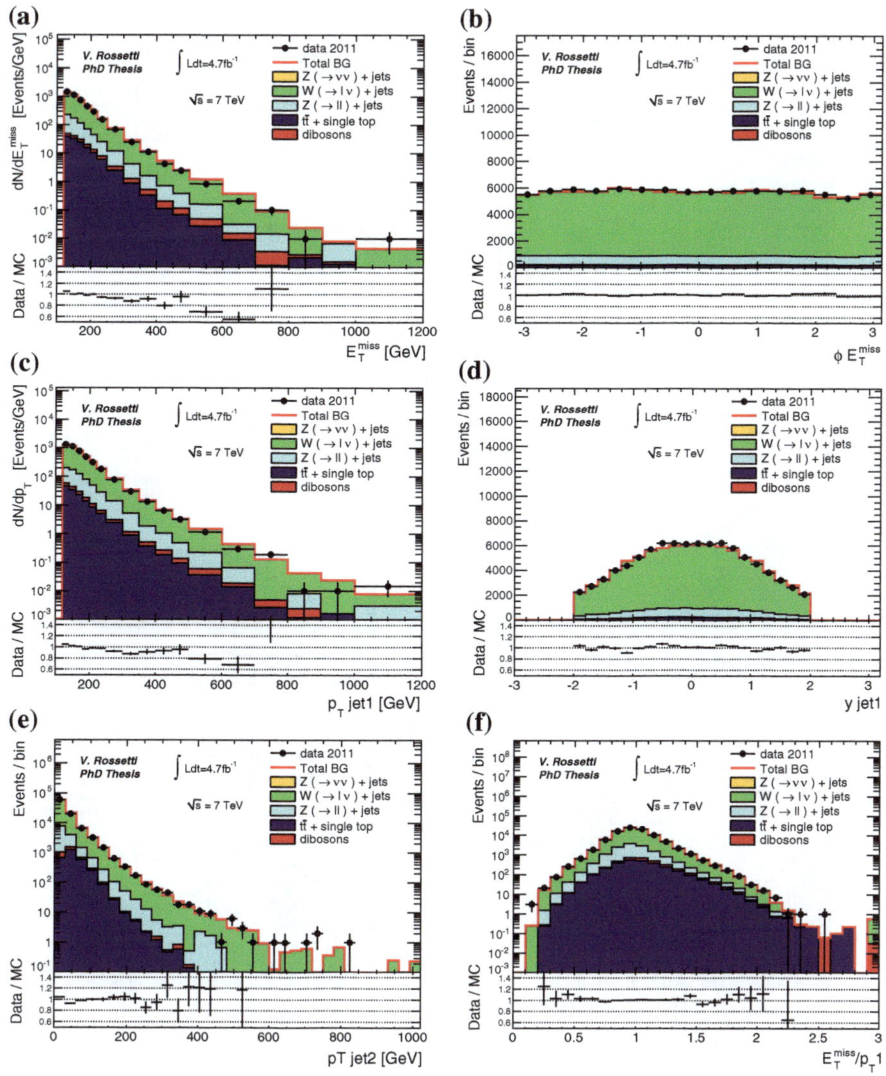

Fig. 3.6 Kinematic distributions of the events in the inclusive muon CR for the selection cuts of region 1 ($E_{\mathrm{T}}^{\mathrm{miss}}$, jet1 $p_{\mathrm{T}} > 120\,\mathrm{GeV}$)

- the energy deposited in the calorimeters by the muons and by the radiated photons (typically few GeV),
- the contribution of the photon propagator and the interference with the Z.

All these differences between SR and CR events are taken into account in the TFs.

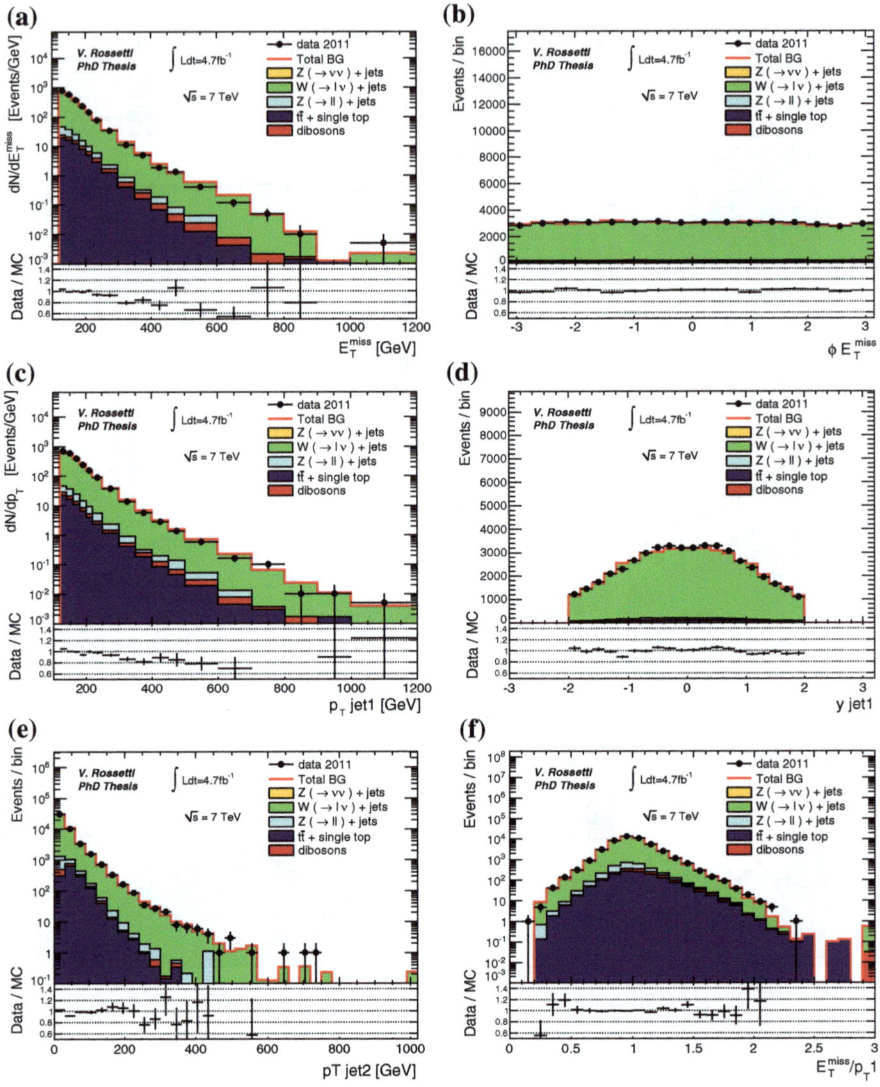

Fig. 3.7 Kinematic distributions of the events in the exclusive $W(\rightarrow \mu\nu) +$jets CR for the selection cuts of region 1 ($E_{\mathrm{T}}^{\mathrm{miss}}$, jet1 $p_{\mathrm{T}} > 120\,\mathrm{GeV}$)

The TF method provides a separate estimation of every Z/W + jets process ($Z(\rightarrow \nu\nu)$, $W(\rightarrow e\nu)$, $W(\rightarrow \mu\nu)$, $W(\rightarrow \tau\nu)$, $Z(\rightarrow \tau\tau)$, and $Z(\rightarrow \mu\mu)$[7]), using one of the four CRs defined previously ($W(\rightarrow \mu\nu)$, $Z(\rightarrow \mu\mu)$, inclusive muon and inclusive

[7]The $Z(\rightarrow ee) +$jets is negligible because of the $E_{\mathrm{T}}^{\mathrm{miss}}$ requirement.

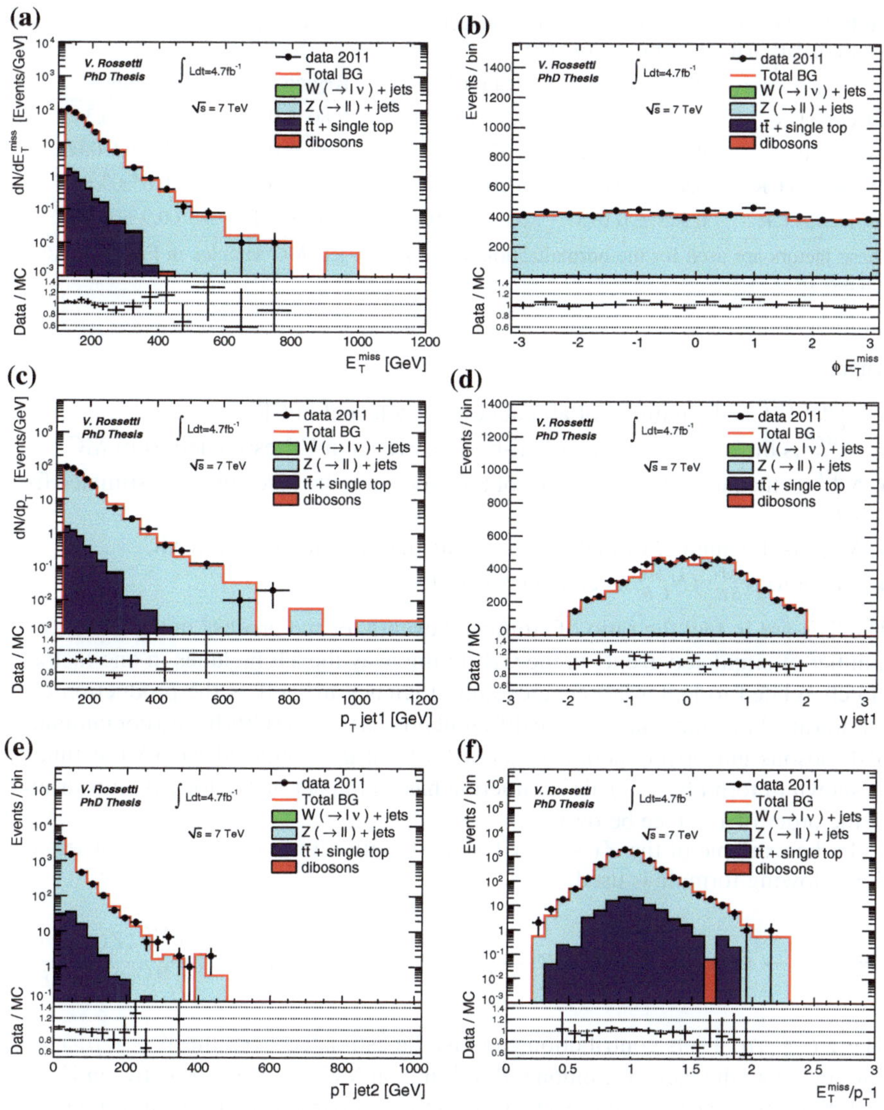

Fig. 3.8 Kinematic distributions of the events in the exclusive $Z(\rightarrow \mu\mu)$ CR for the selection cuts of region 1 ($E_{\mathrm{T}}^{\mathrm{miss}}$, jet1 $p_{\mathrm{T}} > 120\,\mathrm{GeV}$)

electron). For each process a correspondent CR is chosen, and the number of SR events $N_{SR}^{est.}$ is estimated with the formula:

$$N_{SR}^{est.} = (N_{CR}^{data\ TOT} - N_{BG}^{MC}) \times [\frac{N_{SR}^{MC}}{N_{CR}^{MC}}], \tag{3.1}$$

Table 3.2 Summary of the CR normalization factors with their corresponding statistical uncertainties from data and MC in the various CRs

	Region 1	Region 2	Region 3	Region 4
Electron CR	0.926 ± 0.009	0.80 ± 0.03	0.80 ± 0.08	1.0 ± 0.3
Muon CR	0.963 ± 0.005	0.87 ± 0.01	0.81 ± 0.04	0.65 ± 0.09
$W(\rightarrow \mu\nu)$ CR	0.959 ± 0.006	0.85 ± 0.02	0.78 ± 0.05	0.65 ± 0.13
$Z(\rightarrow \mu\mu)$ CR	0.978 ± 0.015	0.89 ± 0.05	1.01 ± 0.17	0.9 ± 0.4

These factors are used for the normalization of the Z/W+jets MC samples in Figs. 3.5, 3.6, 3.7 and 3.8

where:

- $N_{CR}^{data\ TOT}$ is the number of all data events selected in CR
- N_{BG}^{MC} is the number of top and di-bosons events in CR estimated from MC
- N_{SR}^{MC} is the number of SR events of the process under investigation estimated from MC
- N_{CR}^{MC} is the number of Z/W+jets events in CR estimated from MC
- The ratio N_{SR}^{MC}/N_{CR}^{MC} is called transfer factor (TF).

The TFs are in fact the ratio of simulated events for the process in the SR over the number of simulated events in the CR. In this way the TF includes in one factor all effects related to lepton acceptance and efficiency as well as the production cross section and branching ratios of the different processes. The CRs have a contamination of di-bosons and top production (order of 3 %) that are subtracted using MC estimates, as shown in formula 3.1. This subtraction has an impact on the total BG uncertainty of the order of 1 % (see below).

In the example of the $Z(\rightarrow \nu\nu)$+jets estimation from the $Z(\rightarrow \mu\mu)$+jets CR, the following formula is used:

$$N_{SR}^{est.}(Z(\rightarrow \nu\nu)) = (N_{CR}^{data\ TOT}(Z(\rightarrow \mu\mu)) - N_{BG}^{MC}(Z(\rightarrow \mu\mu))) \times [\frac{N_{SR}^{MC}(Z(\rightarrow \nu\nu))}{N_{CR}^{MC}(Z(\rightarrow \mu\mu))}]$$

The $Z(\rightarrow \nu\nu)$+jets contribution can also be estimated from the $W(\rightarrow \mu\nu)$+jets CR, and from the inclusive muon CR, relying on MC for the ratios between Z+jets and W+jets processes. Figure 3.9 presents the comparison of the results for the $Z(\rightarrow \nu\nu)$+jets BG as determined from the different CR. The comparison demonstrates the consistency across different CRs. As expected, the results from the inclusive muon CR are very close to those from the $Z(\rightarrow \mu\mu)$+jets and $W(\rightarrow \mu\nu)$+jets CR. The comparison also illustrates how a data-driven procedure based on the $Z(\rightarrow \mu\mu)$+jets sample alone would suffer from large statistical uncertainties.

With the TF method most of the systematic uncertainties on the BG estimation are largely reduced. For example the luminosity uncertainty does not affect the TF, since is exactly cancelled in the ratio. The jet energy scale (JES) uncertainty is largely reduced as well (see below) since a change on the energy scale would change the denominator and denominator in a very similar way, so that the TF is mostly

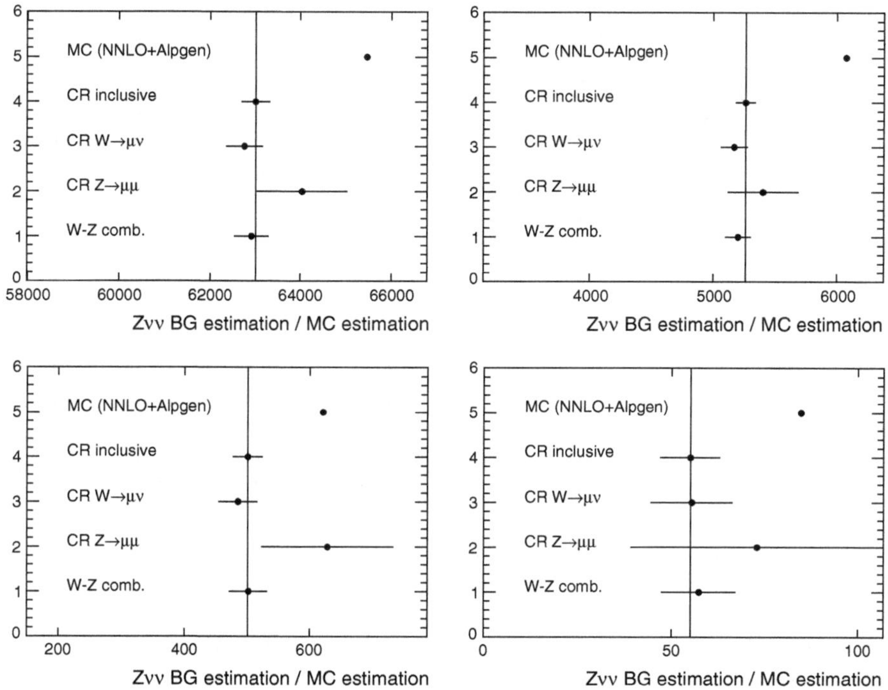

Fig. 3.9 Comparison of $Z(\rightarrow \nu\nu)+$jets estimations using different methods for the four SRs (SR1 and SR2 in the *upper plots*, SR3 and SR4 in the *lower ones*). The *error bands* represent statistical errors only

unchanged. To minimize the effects on the TF it's important that CRs have a final state as close as possible to the SR process that we want to estimate. This argument has driven the choice of the CRs to be used for each SR estimate.

The $Z(\rightarrow \nu\nu)+$jets, $W(\rightarrow \tau\nu)+$jets, $W(\rightarrow \mu\nu)+$jets, and $Z(\rightarrow \mu\mu)+$jets backgrounds are first estimated separately from the $W(\rightarrow \mu\nu)+$jets or $Z(\rightarrow \mu\mu)+$jets CRs. Then, for each process the predictions from the two CRs are combined for the final estimation. The combination is an error-weighted average that takes into account the correlations of the statistical errors. Instead the processes $W(\rightarrow e\nu)+$jets, $Z(\rightarrow \tau\tau)+$jets, and $Z(\rightarrow ee)+$jets are estimated using the inclusive electron CR.

As a cross-check, the $W(\rightarrow \tau\nu)+$jets estimation was carried out using $W(\rightarrow \mu\nu)$, $Z(\rightarrow \mu\mu)$ and inclusive electron CRs and the results were consistent to each other within uncertainties.[8] For the final results, the estimation based on the $W(\rightarrow \mu\nu)$ and $Z(\rightarrow \mu\mu)$ control samples was adopted, because of the higher statistics in these CRs. Nevertheless the use of the inclusive electron sample could also be justified

[8]This comparison consider both the statistical uncertainty and systematic uncertainties that are presented in the following.

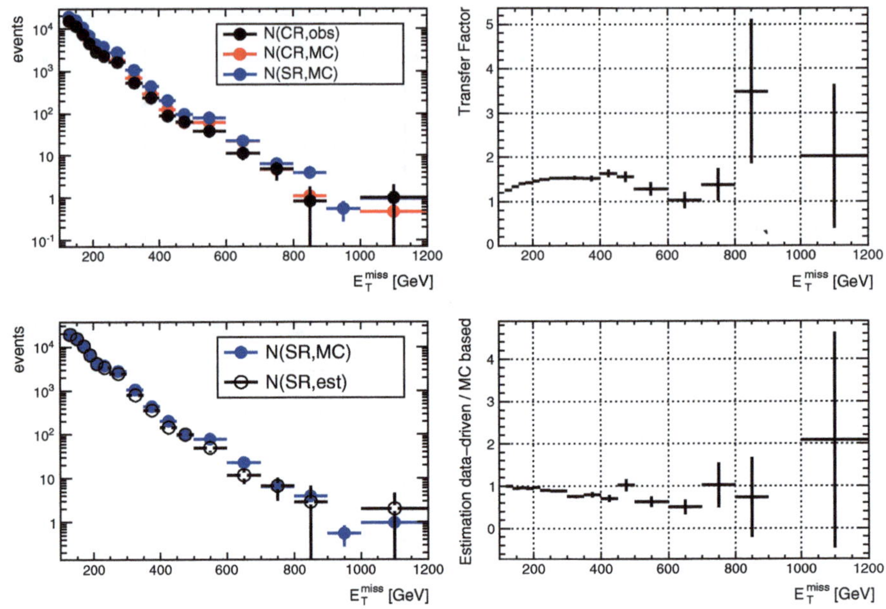

Fig. 3.10 Procedure to build the $Z(\rightarrow \nu\nu)+$jets E_T^{miss} distribution for SR1 from $W(\rightarrow \mu\nu)+$jets CR. *Top left plot* shows the distribution of E_T^{miss} for the data (*black*) and the MC (*red*) in the $W(\rightarrow \mu\nu)+$jets CR and the $Z(\rightarrow \nu\nu)+$jets MC in the SR (*blue*). The *top right plot* shows the transfer factor (TF) as described in the text. The *bottom left plot* shows the estimated $Z(\rightarrow \nu\nu)$ E_T^{miss} distribution (*open circle*) compared to the MC only prediction (*blue*). The *bottom right plot* shows the ratio of the data-driven and the MC-only estimated SR distribution. Only statistical errors are shown

since the τ dominantly decays into a narrow jet leaving energy in the calorimeters, so that $W(\rightarrow \tau\nu)+$jet events behave similarly to $W(\rightarrow e\nu)+$jet events.

The TF method can also be applied bin-by-bin to the different distributions, leading to a data-driven corrected shape in the SR. Technically this means that formula 3.1 is used for the estimation of the number of events in each bin of the distribution. It's worth to mention here that estimating the total number of events from formula 3.1 or by integrating the bin-by bin corrected distributions, give the same result.

Figure 3.10 presents the steps for building the $Z(\rightarrow \nu\nu)+$jets E_T^{miss} distribution in SR1 from the $W(\rightarrow \mu\nu)+$jets CR. The same procedure repeated from the $Z(\rightarrow \mu\mu)+$jets CR is shown in Fig. 3.11.

3.5.1.3 Statistical and Systematic Uncertainties

For each process, the statistical uncertainty on the estimation has been calculated combining the uncertainties from each component of Eq. 3.1: data in CR and MC in both SR and CR. In order to calculate the statistical error on the total Z/W+jets

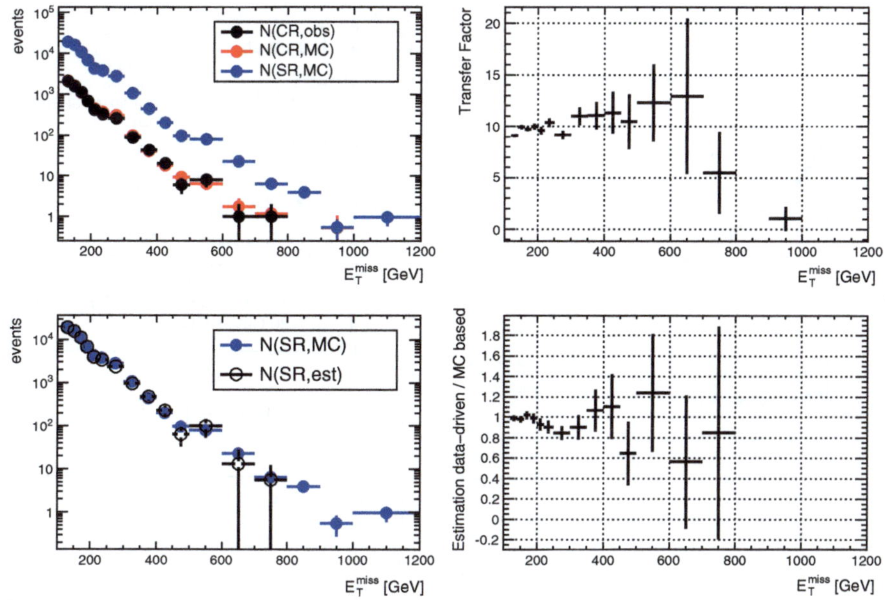

Fig. 3.11 Procedure to build the $Z(\rightarrow \nu\nu)+$jets E_T^{miss} distribution for SR1 from $Z(\rightarrow \mu\mu)+$jets CR. *Top left plot* shows the distribution of E_T^{miss} for the data (*black*) and the MC (*red*) in the $Z(\rightarrow \mu\mu)+$jets CR and the $Z(\rightarrow \nu\nu)+$jets MC in the SR (*blue*). The *top right plot* shows the transfer factor (TF) as described in the text. The *bottom left plot* shows the estimated $Z(\rightarrow \nu\nu)$ E_T^{miss} distribution (*open circle*) compared to the MC only prediction (*blue*). The *bottom right plot* shows the ratio of the data-driven and the MC-only estimated SR distribution. Only statistical errors are shown

BG the uncertainties of all the processes are combined, taking into account the correlations.

Different classes of systematic uncertainty are considered in the Z/W+jets BG determination:

- Z/W+jets MC modeling
- Jet and E_T^{miss} related uncertainties
- Lepton identification
- Background subtraction in control regions
- Trigger efficiency
- Luminosity

Uncertainties on the Z/W+jets MC modeling As we discussed previously, the data-driven method implemented in this BG estimation allows to cancel the leading contribution of the different sources of uncertainty on the MC simulation. In the following, the residual uncertainties on the Z/W+jets MC prediction are assessed. This includes uncertainties related to: parton shower (PS) modeling, matrix element to PS matching, choice of the renormalization and factorization scales, and

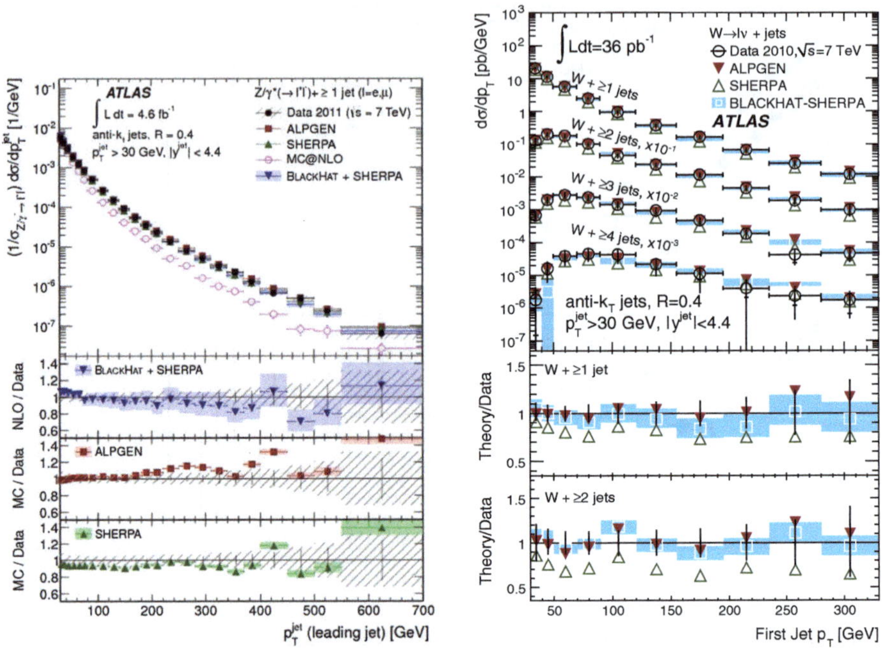

Fig. 3.12 Differential cross section measurement of $Z(\rightarrow \ell\ell)+$jets (*left*) and $W(\rightarrow \ell\nu)+$jets (*right*) as a function of leading jet p_{T}. Pictures are taken from Refs. [31, 32]. The measured differential cross section is compared with predictions from SHERPA and ALPGEN, showing a different behavior of the two MC generator on the jet p_{T} prediction

the choice of the PDF set. The different variations result in different predictions for the processes under study. For example, on the Z + jet and W + jet production ALPGEN and SHERPA give different predictions, in particular on the shape of the jet p_{T}. Figure 3.12a (taken from [31]) and 3.12b (taken from [32]) show the measured differential cross section as a function of the leading jet p_{T} in $Z(\rightarrow \ell\ell)+$jets and $W(\rightarrow \ell\nu)+$jets events, respectively. Considering the systematic uncertainties on these measurements both the ALPGEN and SHERPA predictions are in agreement with the data. Nevertheless, the shape the jet p_{T} in the two MCs has opposite tendency. At $p_{\mathrm{T}} \sim 150\,\mathrm{GeV}$ ALPGEN and SHERPA predictions differ by $\sim$20 % and the difference tends to increase for higher p_{T}.

To assess systematic uncertainties on the MC modeling, the Z/W + jets BG estimations made with ALPGEN and SHERPA have been compared. Because of the large difference between the two MC predictions, this is considered a conservative approach. Two different procedures have been carried out for comparing ALPGEN and SHERPA results. The first compares directly the estimations made with the two sets of MC samples, resulting in SR1 in a difference of 1.6 % with a statistical error of 1.1 %. For SR2, SR3 and SR4 the differences are 3.7 ± 3.0, 8 ± 11 and 4 ± 26 %. Due to the severe lack of statistics in the SHERPA samples the comparison suffers

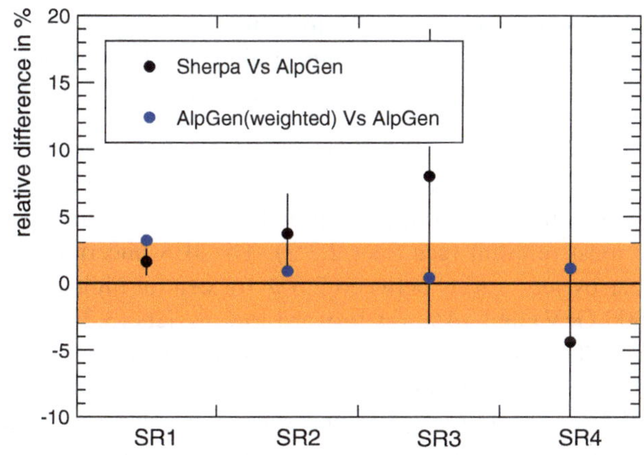

Fig. 3.13 Detail of the comparison between the total Z/W + jets BG using SHERPA and ALPGEN MC samples. *Black points* show the relative difference between the SHERPA and ALPGEN estimations. *Blue points* show the relative change of the ALPGEN results when weights have been applied to correct the boson p_{T} distribution based on the SHERPA prediction. Based on these result a systematic uncertainty of 3 % has been chosen (represented by the *shadowed band*)

from large uncertainties as one increases the $E_{\mathrm{T}}^{\mathrm{miss}}$ and jet p_{T} thresholds. In the second comparison, the ALPGEN events have been re-weighted to follow the p_{T} distribution of the W and Z bosons in SHERPA. The weights are defined by the ratio of the distributions in the two sets of MC samples before any selection is applied. The values of the weights go from 1.15 at low boson p_{T}, to 0.72 for $p_{\mathrm{T}} > 250$ GeV. The comparison between the original ALPGEN predictions and the predictions from the modified samples is about 3 % in SR1 and is smaller than 1 % for the other SRs.

The ALPGEN-SHERPA differences from these two comparisons are summarized in Fig. 3.13. The difference shows no dependency with increased thresholds on p_{T} and $E_{\mathrm{T}}^{\mathrm{miss}}$. Therefore an uncertainty of 3 % has been considered for the four SRs.

Lepton related uncertainties The uncertainty on the lepton identification is not reduced by the data-driven BG estimation, because its effects are not canceled in the TF ratios. Nevertheless, MC events are corrected with lepton identification scale factors to have the same reconstruction efficiency as in data, and the remaining uncertainties are propagated to the final result. The uncertainty on the scale factors are both statistical (from the finite size of the event sample used to estimate them) and systematics (from the method implemented to retrieve them).

In the case of the muon scale factors, this translates into uncertainties between a 0.8 and 3 % on the total BG as the jet p_{T} and $E_{\mathrm{T}}^{\mathrm{miss}}$ increase. Similarly, the uncertainty on the electron scale factors introduces an uncertainty on the total BG that varies between 0.4 and 0.3 %. The effect of the electron systematics is smaller compared to the muon, because the electron CR is used only for the $W(\rightarrow e\nu) +$ jet and $Z(\rightarrow \tau\tau) +$ jet estimations. The uncertainties on the muon momentum scale and resolution have been evaluated and varies between 0.02 % for SR1 and 0.5 % for

SR4. Instead, the uncertainties on the electron energy scale goes from 0.2 % for SR1 and 0.3 % for SR4. The final systematic uncertainty due to the lepton identification (considering both electrons and muons) is 1 and 3 % for SR1 and SR4, respectively.

Jet and $E_{\mathrm{T}}^{\mathrm{miss}}$ related uncertainties Various sources of systematic uncertainties on the jet energy scale (JES) have been considered [24]: calorimeter energy response, MC-based and in-situ calibration, pile-up and presence of close-by jets, partonic flavor, are the most relevant (see Sect. 2.3.3). The JES uncertainty, in absence of pile-up and near-by jets, varies between 2.5 % for central high-p_{T} jets ($|\eta| < 0.8$, $60 < p_{\mathrm{T}} < 800$ GeV) and 14 % for forward low-p_{T} jets ($3.2 < |\eta| < 4.5$, $20 < p_{\mathrm{T}} < 30$ GeV). Close-by jets with $\Delta R < 0.7$ introduce an additional uncertainty of 2–3 %. Multiple interactions in the same bunch-crossings introduce, for jets with $p_{\mathrm{T}} = 30$ GeV ($p_{\mathrm{T}} = 100$ GeV), an uncertainty on the jet momentum of 0.5 % (0.1 %) for each reconstructed vertex. To evaluate the impact on the Z/W + jets BG estimation, the p_{T} of all jets is scaled up and down according to the JES uncertainty, and the estimation is repeated. The impact due to the jet energy resolution (JER) uncertainty has been evaluated by smearing the p_{T} of all the jets according to the its uncertainty. The effect of JER on the total BG estimation is negligible.

The $E_{\mathrm{T}}^{\mathrm{miss}}$ reconstruction has various sources of uncertainties due to the presence of jets, electrons, and soft particles coming from pile-up and underlying event. No uncertainty needs to be estimated regarding the muons since the muons are not part of the $E_{\mathrm{T}}^{\mathrm{miss}}$ reconstruction.

The main $E_{\mathrm{T}}^{\mathrm{miss}}$ uncertainty is related to the JES, and is evaluated varying simultaneously the $E_{\mathrm{T}}^{\mathrm{miss}}$ and the p_{T} of the jets since their uncertainties are correlated. To estimate the impact of the JES on the $E_{\mathrm{T}}^{\mathrm{miss}}$, the relative jet-level variations of the first two leading jets are propagated to the $E_{\mathrm{T}}^{\mathrm{miss}}$. As a cross-check the same procedure has been carried out considering all jets with $p_{\mathrm{T}} > 20$ GeV for the $E_{\mathrm{T}}^{\mathrm{miss}}$ estimator. The first procedure gives larger uncertainties and it has been used for the final results in Table 3.3. The total uncertainty due to $E_{\mathrm{T}}^{\mathrm{miss}}$ and JES on the final number of BG events in the signal region is about 2 % for SR1. It is worth mentioning that the Monte Carlo predictions in the SR and the CR both move by $O(20\,\%)$. Thus the 2 % is the residual uncertainty after applying the TF method. In SR4 the uncertainty increases up to 4 % reflecting, to some extent, the limited MC statistics.

The $E_{\mathrm{T}}^{\mathrm{miss}}$ uncertainty from the electron energy scale has been estimated by varying the electron p_{T} according to its uncertainty, and simultaneously propagating the change to the $E_{\mathrm{T}}^{\mathrm{miss}}$. The effect on the total Z/W + jets estimation is 0.2 % in SR1.

As the $E_{\mathrm{T}}^{\mathrm{miss}}$ is computed using all topo-clusters up to $|\eta| < 4.5$, and not only the ones associated with jets or electrons, the uncertainty from non associated topo-clusters needs to be evaluated. This is done by changing the topo-cluster energy scale according to the E/p studies carried out in ATLAS. The effect of this component on the transfer factors is negligible.

Uncertainties from background subtraction in control regions In the lepton control regions there is a small contribution from the $t\bar{t}$, single top and di-bosons. These processes are subtracted from the CRs using MC estimates. The uncertainty on the subtraction propagates as an uncertainty on the number of data events in the CR, which leads to an uncertainty on the Z/W+jets BG determination in the SR. A conservative 20 % uncertainty is considered on the number of $t\bar{t}$, single top and di-bosons. The uncertainty on these processes is considered fully correlated, because the JES uncertainty (about 16 %) is the dominant uncertainty on these BGs. This subtraction results in about 1 % uncertainty on the Z/W+jets BG in the SRs.

Uncertainty on the luminosity The uncertainty on the absolute integrated luminosity is not considered for the final result since it affects the numerator and the denominator of the TF in the same way, so that the ratio is unchanged. The luminosity uncertainty on top and di-boson production indirectly affects the Z/W+jets estimation. This effect is very small and is considered in the uncertainty on the BG subtraction in the CRs.

Uncertainties on the trigger efficiency As described in Sect. 3.4, the trigger used for this analysis is ∼98 % efficient for SR1 and a ∼1 % difference in efficiency between data and MC has been observed. Instead, for SR2, SR3, and SR4 the trigger selection is fully efficient. In order to evaluate the effect of the trigger turn-on curve, the full Z/W estimation has been repeated with different trigger requirements on MC events. In one case a trigger item with higher thresholds ("EF_xe90_noMu") has been used, and in another case no trigger requirement have been made. This test shows how much the BG estimation can vary when using a very different trigger efficiency in the MC. Differences on the total Z/W+jets BG are 0.2 % at most.

Summary of the uncertainties Table 3.3 summarizes the systematic uncertainties on the total Z/W+jets predictions. The final BG estimation uses the combination of the results from $W(\rightarrow \mu\nu)$ and $Z(\rightarrow \mu\mu)$ CRs to estimate $Z(\rightarrow \nu\nu)$, $W(\rightarrow \tau\nu)$, $W(\rightarrow \mu\nu)$ and $Z(\rightarrow \mu\mu)$ processes.

As stated previously, these processes could be also estimated from the inclusive muon CR. Table 3.3 presents the systematic uncertainties on the BG estimation when predicting these processes from either the inclusive muon CR, or the $W(\rightarrow \mu\nu)$ CR, or the combination of $W(\rightarrow \mu\nu)$ and $Z(\rightarrow \mu\mu)$ CRs. The results show how the systematic uncertainties are at the same level for different definitions of the CRs. Table 3.3 shows also that predictions using the inclusive muon CR have slightly lower systematic uncertainties compared to those from the other CRs, especially for SR1 and SR2. This is explained by the fact that the cuts on M_T in the $W(\rightarrow \mu\nu)$ CR, and on $M_{\mu\mu}$ in the $Z(\rightarrow \mu\mu)$ CR introduce a further difference between SR and CR selections. Previously it has been explained that the systematic uncertainties are largely reduced because they affect the TF numerator and denominator in the same (or similar) way, so that the ratio is much less sensitive to systematic effects. Therefore, if further selection criteria are added in the CRs and not in the SR, the systematic uncertainties are bounded to increase.

Table 3.3 Summary of the systematic uncertainties on the total Z/W + jets background

Systematics	Region 1	Region 2	Region 3	Region 4
Inclusive muon CR				
MC modeling (%)	3.0	3.0	3.0	3.0
$E_{\mathrm{T}}^{\mathrm{miss}}$/JES/JER (%)	1.5	2.7	3.7	4.0
Lepton id./scale/res.	0.8	1.5	1.8	2.7
Background subtraction (%)	1.0	0.8	0.8	1.1
Trigger (%)	0.2	–	–	–
Luminosity	–	–	–	–
Exclusive $W(\rightarrow \mu\nu)$ CR				
MC modeling (%)	3.0	3.0	3.0	3.0
$E_{\mathrm{T}}^{\mathrm{miss}}$/JES/JER (%)	2.2	3.4	3.8	2.6
Lepton id./scale/res. (%)	0.7	1.1	1.4	2.4
Background subtraction (%)	1.1	0.9	1.0	1.2
Trigger (%)	0.2	–	–	–
Luminosity	–	–	–	–
Combined results from $W(\rightarrow \mu\nu)$ and $Z(\rightarrow \mu\mu)$ CRs				
MC modeling (%)	3.0	3.0	3.0	3.0
$E_{\mathrm{T}}^{\mathrm{miss}}$/JES/JER (%)	2.1	2.9	3.9	2.1
Lepton id./scale/res. (%)	0.7	1.1	1.4	2.4
Background subtraction (%)	1.1	0.9	0.9	1.1
Trigger (%)	0.2	–	–	–
Luminosity	–	–	–	–

The final uncertainties are compared for different choices of the muon CRs: the inclusive muon CR, the $W(\rightarrow \mu\nu)$ CR, the combination of both $W(\rightarrow \mu\nu)$ and $Z(\rightarrow \mu\mu)$ CRs

3.5.2 Multi-jet Production

Multi-jet events from a pure QCD interaction enter the mono-jet SRs when one or more jets are mis-measured or completely lost in dead regions of the detector. As illustrated in Fig. 3.14, there are two dominant configurations passing the selection cuts: di-jet and tri-jet events. In both cases, the $E_{\mathrm{T}}^{\mathrm{miss}}$ points to the ϕ direction of the

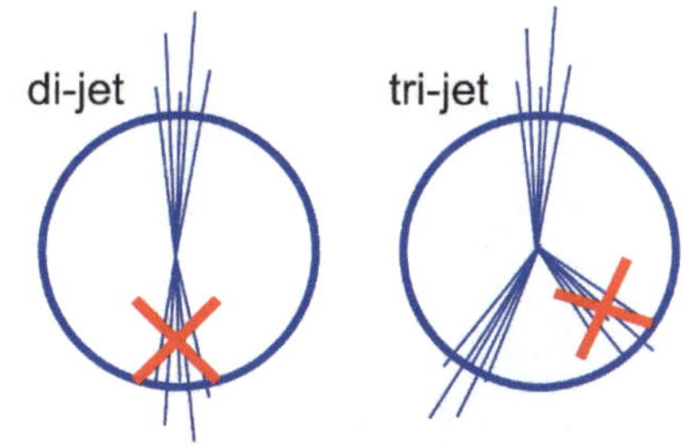

Fig. 3.14 Sketch of the two dominant configurations for the multi-jet events passing the SR selection: di-jet events (*left*) and tri-jet events (*right*)

mis-measured jet. It is much more probable to loose or mis-measure only one jet than two or more jets in the same event. Therefore, multi-jet contamination of events with two or more mis-measured jets can be neglected. This assumption is also supported by dedicated MC studies, that showed that this fraction of events is negligible.

The multi-jet BG is estimated in a data-driven way by extrapolating the p_T spectrum of the mis-measured jet below the 30 GeV threshold, hence in the SR. For each SR, two CRs are defined in order to estimate separately the di-jet and tri-jet BG components.

Di-jet control regions The di-jet CRs are obtained applying the selection of the corresponding SR, but requiring a second jet with $p_T > 30$ GeV and $\Delta\phi(\text{jet2}, E_T^{\text{miss}}) \leq 0.5$. This CR is ortogonal to the SR and is dominated by di-jet events, in which the mis-measurement of the second jet leads to a high E_T^{miss} ($E_T^{\text{miss}} > 120$ GeV). The distribution of $\Delta\phi(\text{jet2}, E_T^{\text{miss}})$ before the cut on this variable is shown in Fig. 3.15a for the E_T^{miss} and jet p_T thresholds corresponding to SR1. For comparison with data, the QCD MC samples are normalized to the observed number of events, A scale

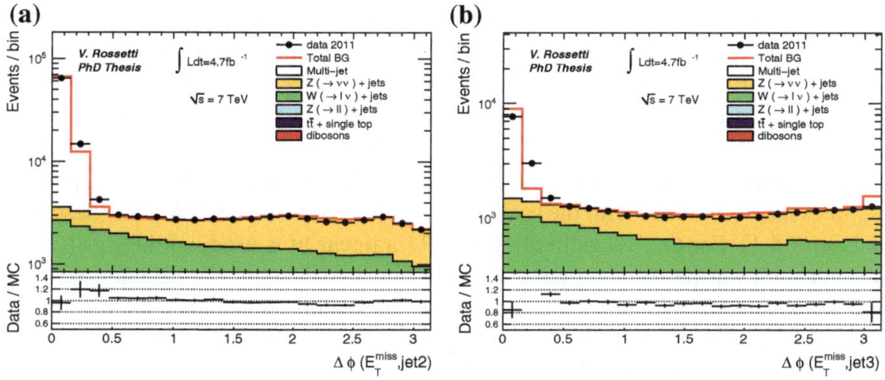

Fig. 3.15 Figure **a** shows the distribution of $\Delta\phi(\text{jet2}, E_T^{\text{miss}})$ in a di-jet configuration. The events with $\Delta\phi \leq 0.5$ are used for the di-jet CR, while the events with $\Delta\phi > 0.5$ are part of the SR, since the second jet is not aligned with the E_T^{miss}. Figure **b** shows $\Delta\phi(\text{jet3}, E_T^{\text{miss}})$ in events with a tri-jet configuration, and the events with $\Delta\phi \leq 0.5$ are used for the tri-jet CR

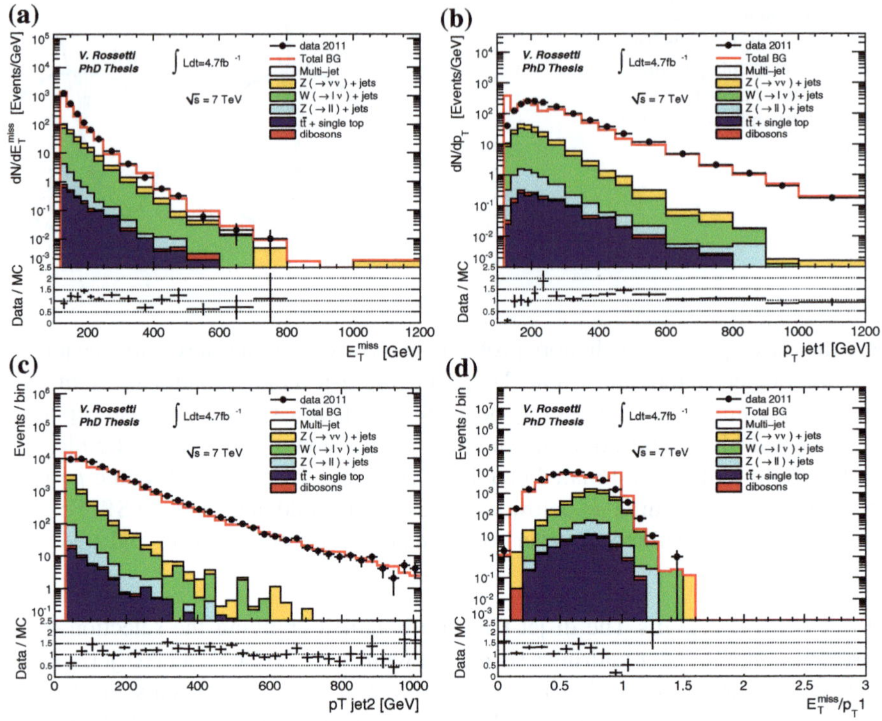

Fig. 3.16 Distribution of $E_{\mathrm{T}}^{\mathrm{miss}}$ (**a**), leading jet p_{T} (**b**), second leading jet p_{T} (**c**), ratio between $E_{\mathrm{T}}^{\mathrm{miss}}$ and leading jet p_{T} (**d**) in the di-jet CR for the first region selection (jet1 p_{T}, $E_{\mathrm{T}}^{\mathrm{miss}} > 120\,\mathrm{GeV}$)

factor of 0.76 is needed to bring the normalization of the MC samples, from the LO cross section provided by PYTHIA, to the number of events in data.

Figure 3.16 shows the distributions of the relevant quantities of the first two leading jets and the $E_{\mathrm{T}}^{\mathrm{miss}}$ for the first di-jet CR (jet1 p_{T}, $E_{\mathrm{T}}^{\mathrm{miss}} > 120\,\mathrm{GeV}$). Figure 3.17 shows the distribution of the second jet p_{T} in the di-jet CRs correspondent to SR1 and SR2 (jet1 p_{T}, $E_{\mathrm{T}}^{\mathrm{miss}} > 120\,\mathrm{GeV}$ and $>220\,\mathrm{GeV}$ respectively), after the non-QCD processes are subtracted from the data. For the subtraction, the Z/W + jets MC samples are normalized with the scale factors retrieved from the Z/W CRs (described in Sect. 3.5.1). For the BG estimation of SR1 a linear fit is used, while for SR2 a fit to a constant was performed giving a conservative estimation. For SR3 and SR4 this BG is considered negligible. The systematic uncertainty is obtained by varying the range of the fit by 10 GeV and by varying up and down the Z/W + jets scale factors by 10 %. Results are shown in Table 3.4.

Tri-jet control regions The tri-jet CRs are obtained applying the selection of the corresponding SR, but requiring a third jet with $p_{\mathrm{T}} > 30\,\mathrm{GeV}$, $\Delta\phi(\mathrm{jet3}, E_{\mathrm{T}}^{\mathrm{miss}}) \leq 0.5$ and no fourth jets with $p_{\mathrm{T}} > 30\,\mathrm{GeV}$. These CRs are orthogonal to the SR and to the di-jet CRs previously defined. They are dominated by tri-jet events, in which

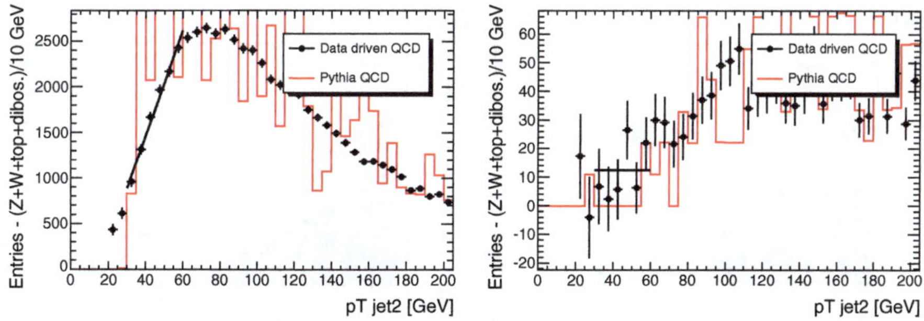

Fig. 3.17 Distribution of the second jet p_T in the di-jet CR with extrapolation below the 30 GeV threshold for SR1 (*left*) and SR2 (*right*). Z/W+jets, top and di-bosons background are subtracted from the data, and data points can therefore be below zero. Note that the relevant part of the distribution is the region below 100 GeV

Table 3.4 Results of the multi-jet BG estimation with statistical and systematic uncertainties

	SR1	SR2	SR3	SR4
Di-jet BG	$750 \pm 30 \pm 640$	$64 \pm 8 \pm 64$	$8 \pm 3 \pm 8$	–
Tri-jet BG	$350 \pm 20 \pm 300$	–	–	–
Total multi-jet BG	$1100 \pm 30 \pm 940$	$64 \pm 8 \pm 64$	$8 \pm 3 \pm 8$	–

The systematic errors from the di-jet and tri-jet component are considered correlated, and therefore summed linearly

the mis-measurement of the third jet leads to a high E_T^{miss} ($E_T^{miss} > 120$ GeV). The distribution of $\Delta\phi(\text{jet3}, E_T^{miss})$ before the cut on this variable is shown in Fig. 3.15b. As for the di-jet CR plots, QCD MC samples are normalized to the data. This leads to a normalization factor for the QCD MC samples of 0.78.

Figure 3.18 shows the distributions of the relevant quantities of the first three leading jets and the E_T^{miss} for the tri-jet CR corresponding to SR1 (jet1 p_T, $E_T^{miss} >$ 120 GeV). Figure 3.19 shows the p_T of the third jet in data and MC, after subtracting all non-QCD processes. For the BG estimation of SR1, a fit with a second-degree polynomial function is used. For SR2, SR3 and SR4 a fit to a constant was performed and the result are compatible with zero. Therefore, this BG is considered negligible for these SRs. The systematic uncertainty is obtained in the same way as for the di-jet estimation. Results are shown in Table 3.4.

Summary of the multi-jet background estimation The BG estimations from the di-jet and tri-jet configurations are summed considering the systematic uncertainties fully correlated. Table 3.4 summarizes the estimations for the four SRs. Although the uncertainty on the multi-jet BG is large O (100 %), its impact on the final result is relatively small since it contributes as a ~1 % of the total BG in SR1 and SR2.

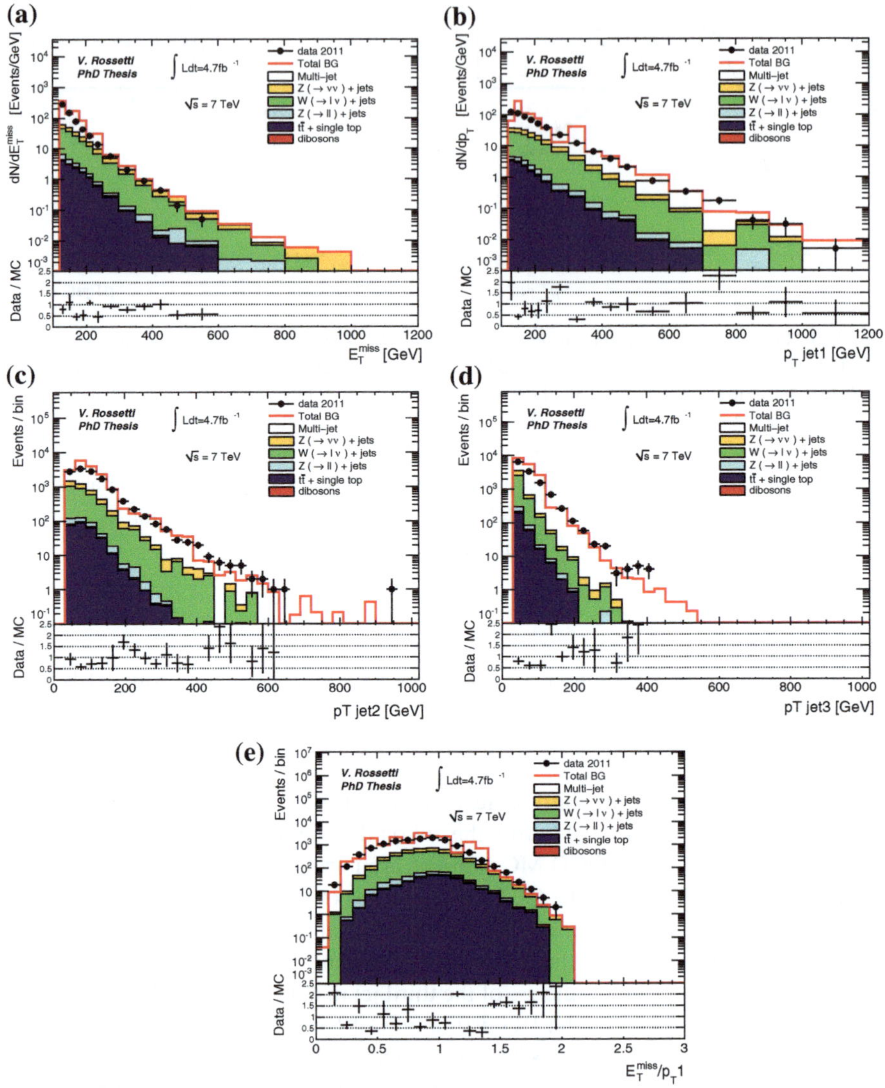

Fig. 3.18 Distribution of $E_{\mathrm{T}}^{\mathrm{miss}}$ (**a**), leading jet p_{T} (**b**), second leading jet p_{T} (**c**), third leading jet p_{T} (**d**), ratio between $E_{\mathrm{T}}^{\mathrm{miss}}$ and leading jet p_{T} (**e**) in the tri-jet CR for the first region selection (jet1 p_{T}, $E_{\mathrm{T}}^{\mathrm{miss}} > 120\,\mathrm{GeV}$)

This QCD multi-jet estimation focuses on the expected number of BG events, but does not predict the shape of the $E_{\mathrm{T}}^{\mathrm{miss}}$ and jet distributions. PYTHIA MC events are used in order to build distributions of the multi-jet BG in each of the SR plots. The integral of the distributions is normalized to the expected number of events in each SR.

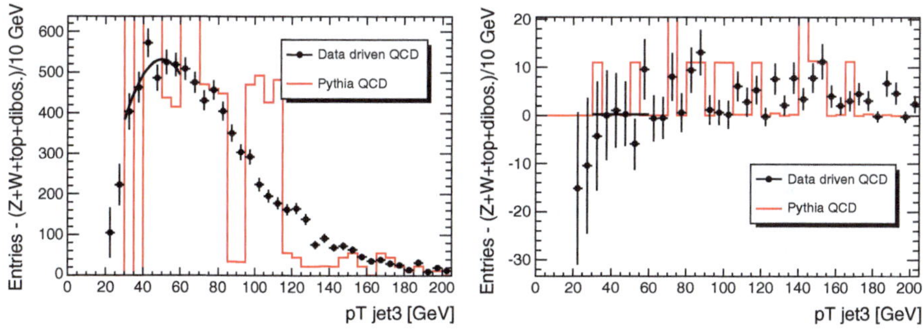

Fig. 3.19 Distribution of the third jet p_T in the tri-jet CR with extrapolation below the 30 GeV threshold for SR1 (*left*) and SR2 (*right*). For region 2 the result of the extrapolation is compatible with zero. Therefore this contribution is considered negligible. Z/W + jets, top and di-bosons background are subtracted from the data, and data points can therefore be below zero. Note that the relevant part of the distribution is the region below 100 GeV

3.5.3 Other SM Processes

The BG estimation for single top, $t\bar{t}$ and di-boson production is based on pure MC predictions. The MC samples are normalized to NLO cross sections calculated with MC@NLO for single top, $t\bar{t}$, and calculated with MCFM [17, 33] for di-boson production. Top and di-boson production represent a fraction of the total BG that goes from 1.3 % for SR1 to 2 % to SR4 (see Table 3.6).

Systematic uncertainties on E_T^{miss}/JES and lepton identification have been estimated in the same way described in Sect. 3.5.1 for the Z/W + jet BG. The luminosity uncertainty (3.4 %) is added in quadrature. The total uncertainty for both top and di-boson production goes from 15 % in SR1 to 18 % in SR4, dominated by E_T^{miss}/JES. A conservative uncertainty of 20 % has been adopted for all SRs.

3.5.4 Non-collision Background

Some of the events passing the mono-jet kinematic selection are characterized by beam-halo particles overlaid with genuine low energy collisions. These particles, mainly muons, are generated from protons traveling in the direction of the experiment and hitting either the LHC collimation system or gas molecules in the beam-pipe or the beam-pipe itself. Figure 3.20 shows an event display in which a halo muon travels almost parallel to the beam-pipe and leaves a large deposit of energy in one region of the calorimeter. This kind of events can be selected as a mono-jet event. In order to reject these events, special cleaning requirements have been implemented (as explained in Sect. 3.4). However, few of these events still pass the selection and constitute what is called the "non-collision" BG. See Ref. [34] for an extensive review on this BG.

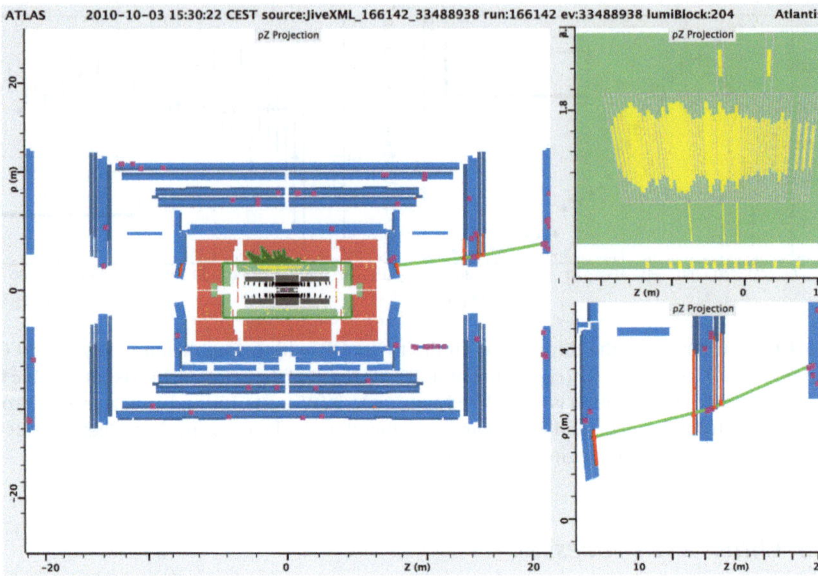

Fig. 3.20 Event display of a typical non-collision event that can pass in the mono-jet selection. A halo muon travelling almost parallel to the beam axis leaves a substantial energy deposit in the LAr calorimeter. The *right side* of the figure shows a detail of the calorimeter deposit and the hits of the outgoing muon in the Muon Spectrometer. This figure as been taken from [34]

A dedicated algorithm has been used in order to estimate the remaining non-collision BG. The algorithm combines calorimeter clusters of energy with hits on both A and C sides of the Muon Spectrometer, matching them in ϕ. In the following, this method will be referenced as the "two-sided" algorithm and it will be used for the baseline BG estimation. A similar algorithm, called "one-sided", requires a matching on only one side of the Muon Spectrometer, and it will be used as a crosscheck. In both methods the timing information from the different sub-detectors is required to be consistent with a muon travelling through the detector along the beam-pipe. More details of these algorithms are given in [34].

For a given SR the number of non-collision events can be estimated as:

$$N_{\text{non-coll.}} = \frac{N_{\text{halo}}}{\epsilon_{tag}^{\text{non-coll.}}},$$

where N_{halo} is the number of SR events tagged by the algorithm, and $\epsilon_{tag}^{\text{non-coll.}}$ is the efficiency to tag non-collision events.

To measure the identification efficiency $\epsilon_{tag}^{\text{non-coll.}}$, a sample of events from unpaired bunch crossings has been selected with the mono-jet kinematic selection. This sample is composed by non-collision events since one of the two proton beams is empty, and can be considered as a control region. The measured identification efficiency

Table 3.5 Number of non-collision background events in the four SRs as estimated by the "two-sided" and "one-sided" methods

SR	Two-sided method		One-sided method	
	N_{halo}	$N_{\text{non-coll.}} \pm$ stat. $\pm$ sys.	N_{halo}	$N_{\text{non-coll.}} \pm$ stat. $\pm$ sys.
1	92	$580 \pm 60 \pm 60$	121	$610 \pm 60 \pm 60$
2	4	$25 \pm 13 \pm 3$	5	$25 \pm 11 \pm 3$
3	0	–	0	–
4	0	–	0	–

The "two-sided" estimates have been taken as default for the final results, while the ones from the "one sided" method are used as cross-check

for the"two-sided" method is $\epsilon_{tag}^{\text{non-coll.}} = 16\,\%$, while the one from the "one-sided" method is $\epsilon_{tag}^{\text{non-coll.}} = 20\,\%$.

In this control sample there is a special class of events in which an unpaired bunch crossing with an empty bunch in one beam is followed by another unpaired bunch crossing with an empty bunch in the other beam, separated by 25 ns. This class of events may lead to double-counting of the events as some reconstructed events are observed with times shifted by 25 ns, i.e. they belong to the neighboring bunch crossing. The amount of these kind of events in the control sample is around 10 %. Since no dedicated studies have been made on these events a 10 % of relative systematic uncertainty is considered on the tagging efficiency.

Table 3.5 summarizes the results from the one-sided and two-sided methods. The difference between the two estimates in SR 1 is ~5 %, that is well within the systematic uncertainty. It has to be noticed that the non-collision BG contribution is very small, as it represents a ~0.5 and a ~0.3 % of the total BG estimate for SR 1 and 2 respectively. For region 3 and 4 no events have been tagged by the two algorithms, so this BG is considered negligible.

As for the QCD multi-jet BG, this estimate focuses on the expected number of BG events, and does not predict the shape of the $E_{\text{T}}^{\text{miss}}$ and jet distributions. For the non-collision BG the distributions have been estimated from data events tagged by the beam-halo tool. The integral of the distributions has been normalized to the expected number of events in each SR.

3.6 Results

In this section, the data in the SRs are compared with the SM predictions for the BG. Figures 3.21 and 3.22 present the distributions of the $E_{\text{T}}^{\text{miss}}$ and leading jet p_{T} in the SR1. The uncertainties showed in the figures take into account only the statistical uncertainties from data and MC. A good agreement between data and SM predictions is found. χ^2 tests performed on the two distributions of Figs. 3.21 and 3.22 (considering only statistical uncertainties) lead to χ^2 per degree of freedom

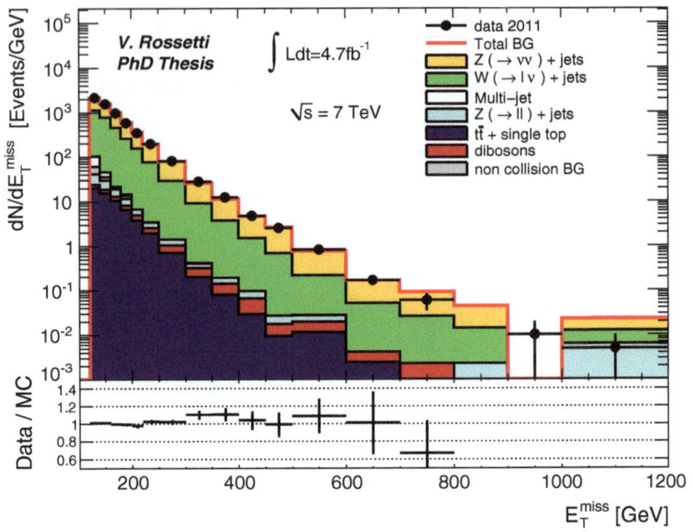

Fig. 3.21 $E_{\mathrm{T}}^{\mathrm{miss}}$ distribution of data and estimated BG in the signal region SR1. The *error bands* in the ratio plot reflect only the statistical uncertainty

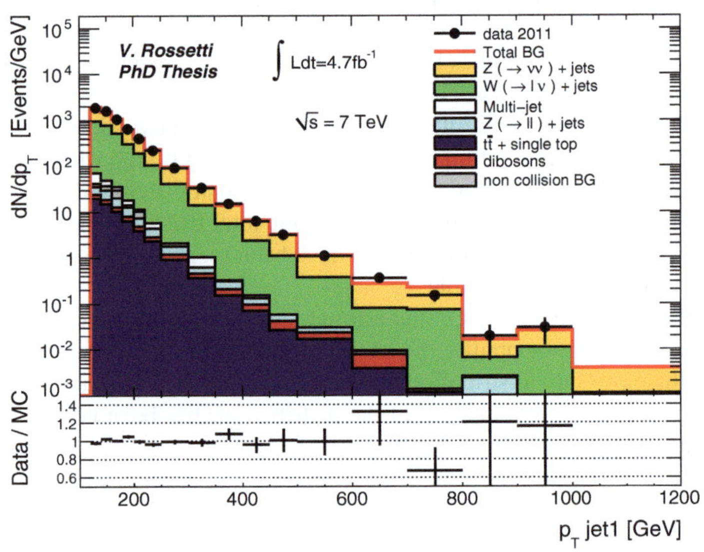

Fig. 3.22 Leading jet p_{T} distribution of data and estimated BG in the signal region SR1. The *error bands* in the ratio plot reflect only the statistical uncertainty

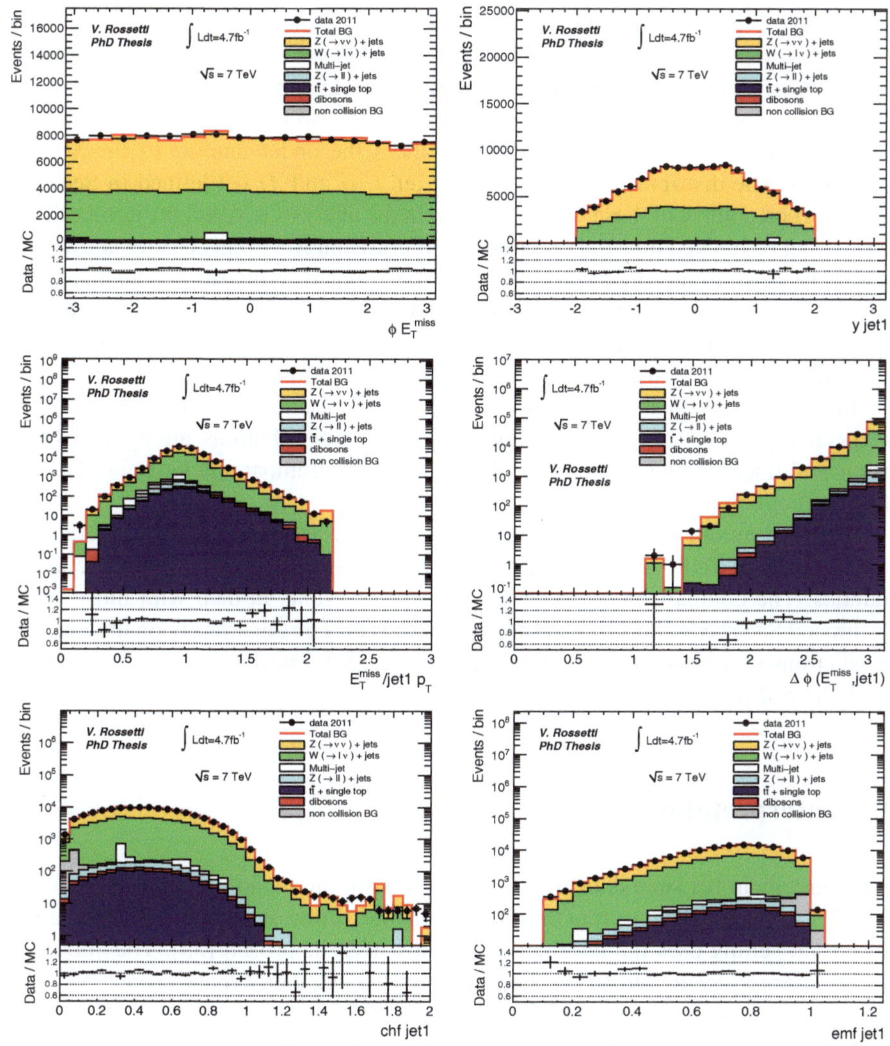

Fig. 3.23 Kinematic distributions of data and estimated BG in the signal region SR1. The *error bands* in the ratio plot reflect only the statistical uncertainty

values 1.2 and 1.1 respectively. This shows a good understanding of the Z/W + jets processes, in a large range of E_T^{miss} and leading jet p_T.

Figure 3.23 shows other quantities related to the E_T^{miss} and leading jet in the SR1. The E_T^{miss} azimuthal angle shows a relatively flat distribution, and a good agreement is found between data and estimated BG. The effect of the LAr hole (see Sect. 3.1) is visible both at $\phi \sim -0.7$ (where the E_T^{miss} is pointing to the LAr hole) and at $\phi \sim 2.5$ (where the leading jet points to it), and is well reproduced by the BG estimation. Figure 3.23 shows also the ratio between the $E_{\mathrm{T}}^{\mathrm{miss}}$ and leading jet p_{T}. Configurations in which $E_{\mathrm{T}}^{\mathrm{miss}}$ and leading jet p_{T} do not fully balance are caused by either the misreconstruction of one of the two objects or by additional radiation and the presence of subleading jets. A good agreement between data and SM predictions is observed, giving additional confidence about the understanding of the SM BG processes. The distributions of the leading jet f_{CH} and f_{EM} (defined in Sect. 3.4) show a good control of these variables, used in the analysis to reject the non-collision BG. It can be noticed how the remaining non-collision BG events cluster at low f_{CH} and high f_{EM}. Figure 3.24 shows distributions related to the second leading jet in SR1. The very good agreement between data and SM predictions testifies a good description of the jet radiation, the p_{T} spectrum, the jet angular distributions, and the ratio between the transverse momenta of two leading jets.

Figures 3.25, 3.26 and 3.27 show instead the distributions for SR2, SR3 and SR4 respectively. A good agreement is found also in these distributions, altought the larger statistical uncertainties make more difficult a shape comparison between data and SM predictions.

The total number of events in the SRs and the corresponding SM BG predictions are reported in Table 3.6. The latter includes both statistical and systematic uncertainties. The comparison between the number of events in data with those of the estimated SM predictions is also shown in Fig. 3.28. The uncertainty on the total SM predictions vary between 4.0 % in SR1 to 17 % in SR4. The data is in good agreement with the SM predictions.

3.7 Interpretations

The agreement between data and SM prediction in the mono-jet analysis, based on 4.7 fb^{-1} of 7 TeV data, is translated into exclusion limits on physics beyond the SM. The limit setting uses the modified frequentist CL_S approach [35]. The "observed limits" are retrieved comparing the probability of the data events in each SR to be compatible with the predicted BG, or with the BG plus a given signal. The probabilities are computed from poissonian distributions with the mean set to the mean value of the BG and signal predictions. To reproduce these probabilities, a number of pseudo-experiments (typically 10^4–10^5) is employed. Systematic uncertainties are treated as nuisance parameters, and their correlations in the signal and BG predictions are taken into account. The "expected limits" are computed with the same procedure but in the hypothesis of observing in data the nominal estimated number of BG events.

Table 3.7 lists the 95 % confidence level (CL) upper limits on the visible cross section σ_{vis} of any new process entering in the mono-jet selection. These bounds are called model independent limits. The visible cross section is defined as $\sigma_{vis} = \sigma \times A \times \epsilon$, where σ is the production cross section, A is the acceptance of the selection (without considering detector effects), and ϵ is the experimental efficiency to select the signal events. Here, the only systematic uncertainty considered on

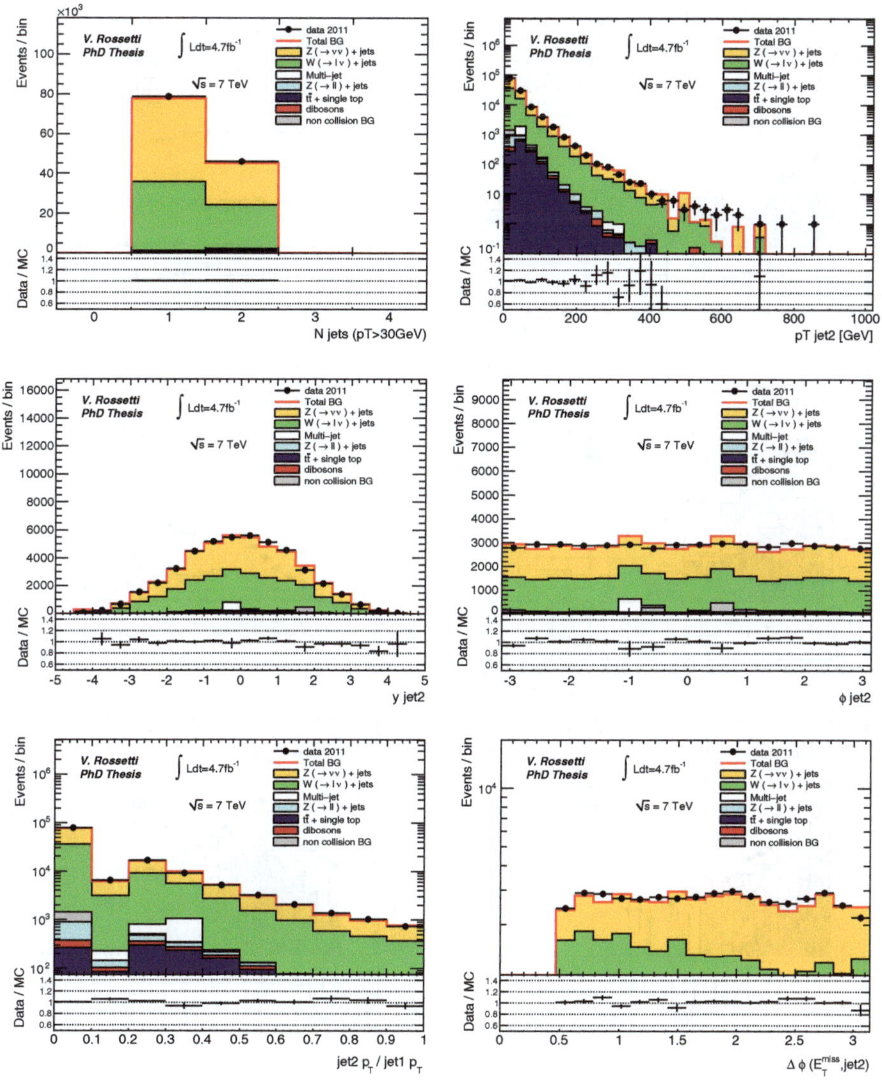

Fig. 3.24 Kinematic distributions of data and estimated BG in the signal region SR1. The *error bands* in the ratio plot reflect only the statistical uncertainty

the hypothetical signal is the luminosity uncertainty. Table 3.7 presents also results corresponding to the 90 % CL limits in order to facilitate the comparison with other direct and indirect Dark Matter searches (see Sect. 1.4.2).

A detailed study using MC simulated events is performed to calculate the efficiency ϵ. This quantity allows to estimate the impact of the model independent limits on any new physics signal, without having to consider the detector description. The

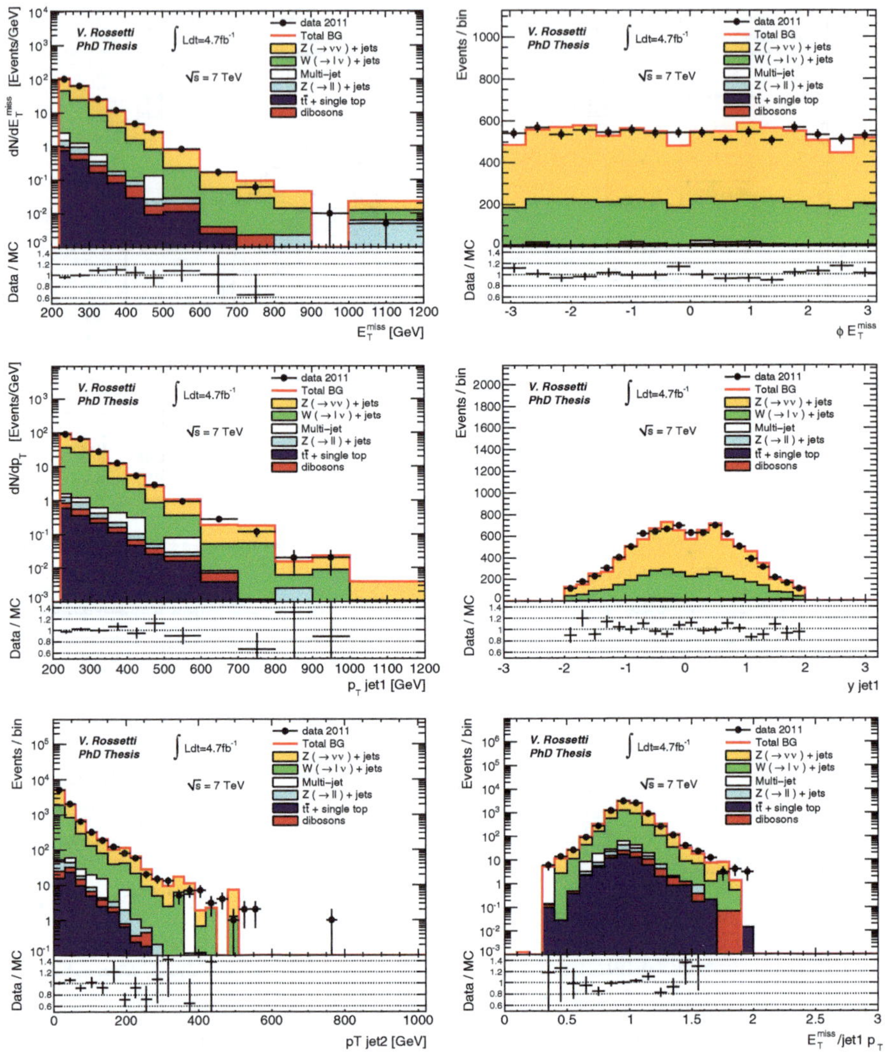

Fig. 3.25 Kinematic distributions of data and estimated BG in the signal region SR2. The *error bands* in the ratio plot reflect only the statistical uncertainty

efficiency is calculated as the ratio of event yields when the mono-jet selection is applied considering the full simulation of the detector, and when is applied on particle-level quantities. For the latter, anti-kt jets with R = 0.4 built with final state particles as constituents, are used instead of the calorimeter-based jets. The particle-level $E_{\mathrm{T}}^{\mathrm{miss}}$ has been defined from the four-momenta of all final-state muons, neutri-

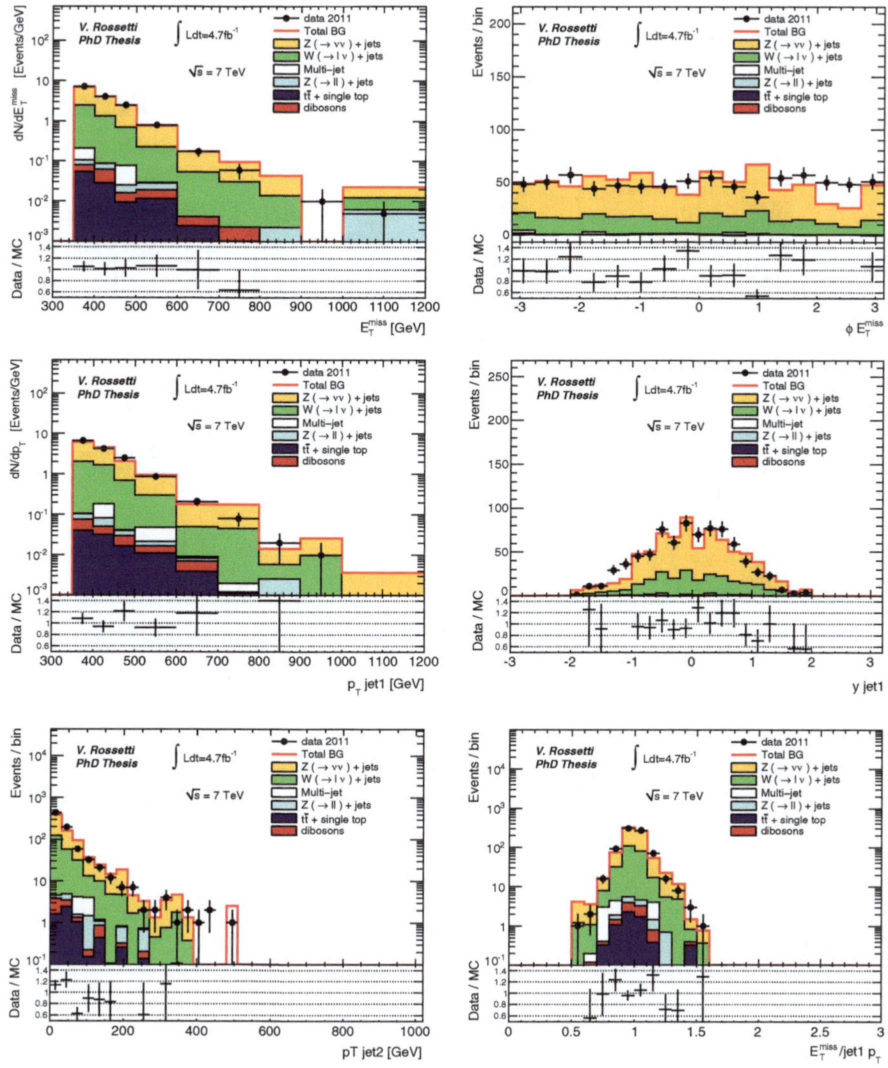

Fig. 3.26 Kinematic distributions of data and estimated BG in the signal region SR3. The *error bands* in the ratio plot reflect only the statistical uncertainty

nos and new invisible particles. Different MC samples have been used to estimate ϵ for different processes: $Z(\rightarrow \nu\nu) + \text{jet}$, WIMPs and ADD signals. The typical efficiencies ϵ are ∼83 %, approximately independent of the process under study.

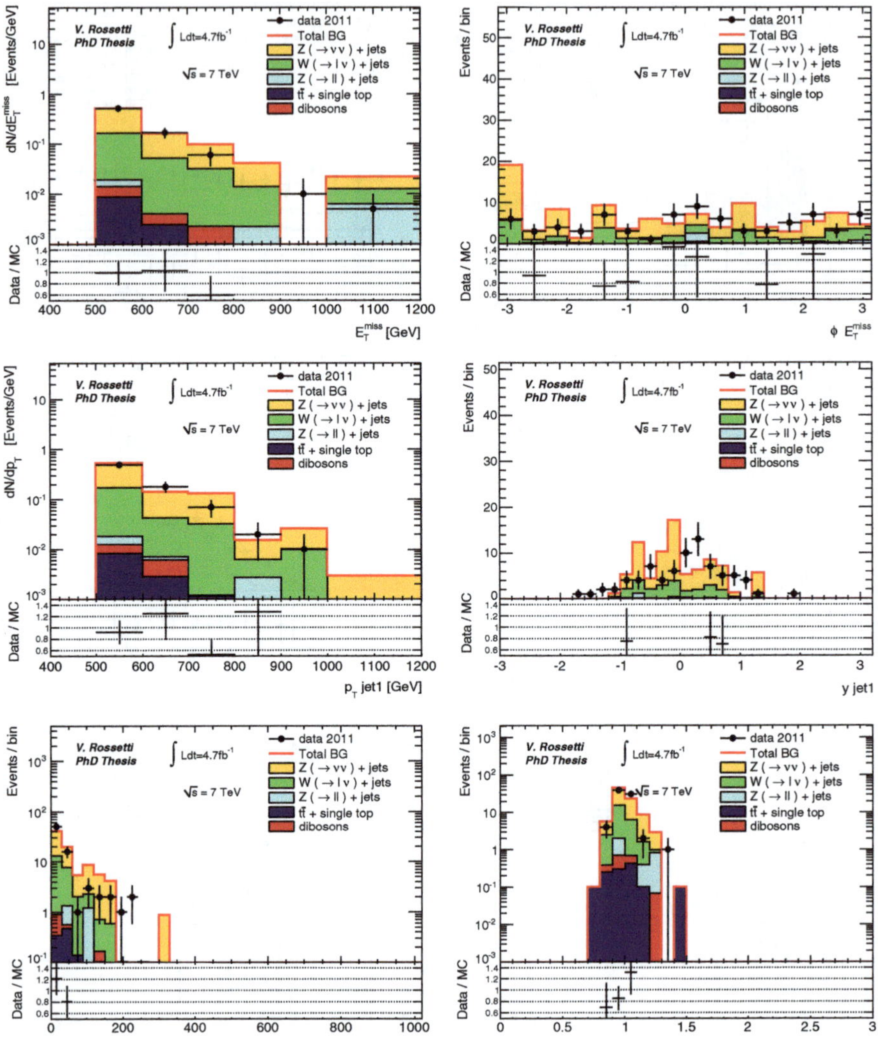

Fig. 3.27 Kinematic distributions of data and estimated BG in the signal region SR4. The *error bands* in the ratio plot reflect only the statistical uncertainty

3.7.1 Limits on Graviton Production in the ADD Model

In this section, the results of the mono-jet analysis are interpreted in the context of the large extra dimensions scenario proposed in Ref. [36] (see Sect. 1.4.1). In this model, new compactified extra dimensions are added to the space-time, with only the graviton field being allowed to propagate through the new extra spatial dimensions.

Table 3.6 Summary of background estimations and total number of observed events in 4.7 fb^{-1} of 7 TeV data for the four mono-jet signal regions

	Background predictions ± (stat. data) ± (stat. MC) ± (syst.)			
	Signal region 1	Signal region 2	Signal region 3	Signal region 4
$Z(\to \nu\nu)+$jets (from $W(\to \mu\nu)$ CR)	$62800 \pm 300 \pm 300 \pm 3000$	$5170 \pm 90 \pm 60 \pm 250$	$490 \pm 30 \pm 20 \pm 30$	$55 \pm 9 \pm 6 \pm 2$
$Z(\to \nu\nu)+$jets (from $Z(\to \mu\mu)$ CR)	$64000 \pm 800 \pm 600 \pm 2400$	$5400 \pm 200 \pm 200 \pm 200$	$630 \pm 80 \pm 60 \pm 40$	$73 \pm 28 \pm 20 \pm 5$
$Z(\to \nu\nu)+$jets (W-Z comb.)	$62900 \pm 300 \pm 300 \pm 2900$	$5200 \pm 80 \pm 60 \pm 230$	$500 \pm 30 \pm 20 \pm 30$	$57 \pm 8 \pm 6 \pm 2$
$W(\to \tau\nu)+$jets	$31200 \pm 100 \pm 200 \pm 1300$	$1780 \pm 30 \pm 30 \pm 90$	$133 \pm 7 \pm 6 \pm 5$	$13 \pm 2 \pm 2 \pm 2$
$W(\to e\nu)+$jets	$13900 \pm 100 \pm 100 \pm 600$	$690 \pm 20 \pm 20 \pm 30$	$47 \pm 4 \pm 3 \pm 3$	$5.4 \pm 1.4 \pm 1.2 \pm 0.5$
$W(\to \mu\nu)+$jets	$11380 \pm 50 \pm 100 \pm 520$	$690 \pm 10 \pm 20 \pm 50$	$53 \pm 3 \pm 4 \pm 6$	$6 \pm 1 \pm 1 \pm 1$
$Z(\to \tau\tau)+$jets	$480 \pm 4 \pm 10 \pm 17$	$20 \pm 1 \pm 2 \pm 2$	$2.4 \pm 0.2 \pm 0.7 \pm 0.5$	$0.7 \pm 0.2 \pm 0.6 \pm 0.3$
$Z(\to \mu\mu)+$jets	$357 \pm 2 \pm 15 \pm 13$	$22 \pm 0 \pm 3 \pm 2$	$2.1 \pm 0.1 \pm 0.8 \pm 0.1$	$0.6 \pm 0.1 \pm 0.4 \pm 0.1$
$Z(\to ee)+$jets	$0.5 \pm 0.5 \pm 0.1$	–	–	–
Multi-jets	$1110 \pm 30 \pm 940$	$64 \pm 8 \pm 64$	$8 \pm 3 \pm 8$	–
$t\bar{t}$ + single t	$1260 \pm 10 \pm 250$	$59 \pm 2 \pm 12$	$6 \pm 1 \pm 1$	$1.3 \pm 0.3 \pm 0.3$
Di-bosons	$289 \pm 3 \pm 58$	$27 \pm 1 \pm 5$	$4.3 \pm 0.4 \pm 0.9$	$0.9 \pm 0.2 \pm 0.2$
Non-collision background	$580 \pm 60 \pm 60$	$25 \pm 13 \pm 3$	–	–
Total background	$123500 \pm 500 \pm 500 \pm 4900$	$8600 \pm 100 \pm 100 \pm 400$	$760 \pm 40 \pm 20 \pm 40$	$84 \pm 11 \pm 8 \pm 4$
Data	124,724	8632	785	77

In cases where both data and MC play a role in the statistical errors, the two contributions are shown separated as second and third uncertainties. The last quoted uncertainty is the systematic uncertainty

This results in a greatly reduced strength of gravity, whose real scale M_D could be closer to the electroweak scale. The agreement between data and SM predictions described in the previous section is translated into 95 % CL limits on the parameters of this model, using both LO and NLO pQCD predictions for the signal.

Figure 3.29 shows the $E_{\mathrm{T}}^{\mathrm{miss}}$ distributions for $Z(\to \nu\nu)+$jet and ADD graviton production, after the selection of SR1 is required. The plot shows that the $Z(\to \nu\nu)+$jet distribution is steeper compared to the ADD graviton. This indicates that a harder cut on $E_{\mathrm{T}}^{\mathrm{miss}}$ and leading jet p_{T} will increase the sensitivity to the signal, provided that the uncertainties on the BG estimation are kept under control.

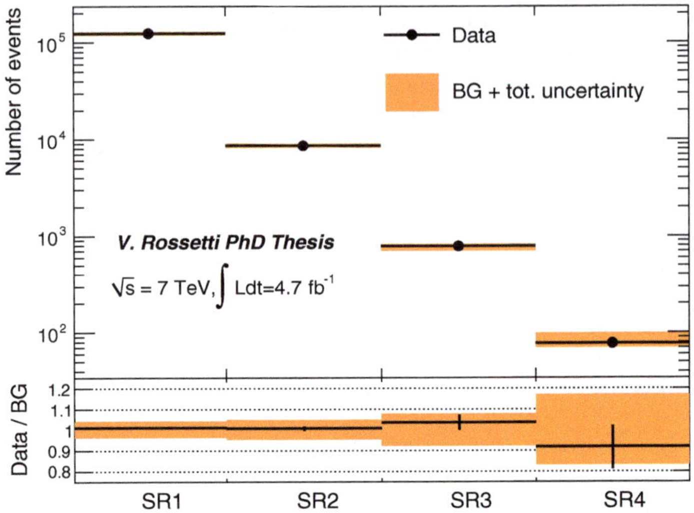

Fig. 3.28 Comparison of the number of observed events in 4.7 fb^{-1} of 7 TeV data to the expected number of BG events in the four signal regions. The *shadowed error band* shows the total uncertainty of the BG estimation

Table 3.7 Model independent limits on the visible cross section on any beyond the Standard Model process for the four SRs

Model independent limits on σ_{vis} (pb)				
	SR1	SR2	SR3	SR4
σ_{vis}^{obs} at 90 % CL	1.63	0.13	0.026	0.0055
σ_{vis}^{exp} at 90 % CL	1.54	0.15	0.020	0.0064
σ_{vis}^{obs} at 95 % CL	1.92	0.17	0.030	0.0069
σ_{vis}^{exp} at 95 % CL	1.82	0.18	0.024	0.0079

Systematic uncertainties Systematic uncertainties on the LO signal yield have been evaluated and are divided into experimental and theoretical ones. The first category includes the uncertainties on jet and $E_{\mathrm{T}}^{\mathrm{miss}}$ scales and resolution, total integrated luminosity, and trigger efficiency. The theoretical category refers to uncertainties on the renormalization and factorization scales, the choice of the PDFs, and the modeling of the initial and final state radiation (ISR/FSR). It is important to mention that the ADD scale M_D impacts the signal yield only through the cross section, and its effect on the signal acceptance is essentially negligible. For this reason, the systematic uncertainties are presented here only as a function of the number of extra-dimensions n.

The dominant experimental uncertainty is the one on the $E_{\mathrm{T}}^{\mathrm{miss}}$/JES and JER, that has been evaluated following the same procedure as in the case of the Z/W+jets BG (see Sect. 3.5.1). Signal yields vary between 2.5 and 12 % depending on the SR

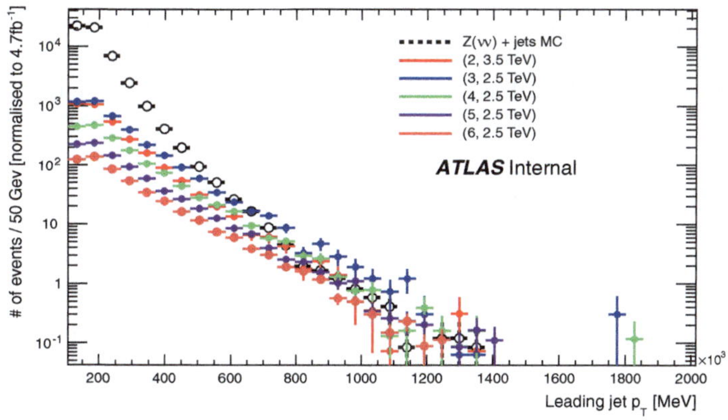

Fig. 3.29 E_T^{miss} distributions for $Z(\rightarrow \nu\nu)+$jet and ADD signal normalized to 4.7 fb^{-1}. Signals are shown for different number of extra dimensions n and for different values of the gravity scale M$_D$. The figure is taken from Ref. [30]

and on n. This uncertainty is smaller compared to the one on Z/W + jet (typically between 10 and 20 %), because the latter is characterized by a steeper shape of the E_T^{miss}. The uncertainty on the luminosity introduces an additional 3.9 % uncertainty in the four SRs, while the trigger efficiency results in a 1 % uncertainty only for SR1.

The uncertainty due to PDFs are evaluated using the Hessian method [37] with the 44 PDF error sets associated with CTEQ6.6.[9] The PDF uncertainty has an impact on the signal yields between 4 and 14 % depending on n, affecting mainly the cross sections, rather than the acceptances.

The uncertainty on the modeling of the ISR/FSR has been evaluated varying the parameters that regulate the parton shower in a range that is consistent with the experimental data. This uncertainty varies the signal acceptance between 3 and 14 % depending on n and on the SR considered.

The uncertainty on the factorization and renormalization scales is the dominant theoretical uncertainty and affects mainly the cross section leaving the acceptance essentially unchanged. Variations of the scales by factors two and one-half, result in changes of 25–35 % on the signal yield, increasing with increasing E_T^{miss}, as it is typically expected for a LO calculation. In the case of NLO predictions, this uncertainty is reduced to 10 %. Table 3.8 summarizes the impact of all the systematic uncertainties on the signal yields in the four SRs and for different n.

Exclusion limits on M_D and on the graviton + jets production cross section 95 % CL lower limits on M$_D$ are computed independently for each SR. Exclusion bounds from SR4 give the best expected limit on and therefore are chosen for the final results.

[9]The 44 error sets are associated with a 90 % CL, when interpreted in terms of gaussian errors. The resulting uncertainty has been divided by a factor 1.645 to translate the 90 % CL error into a 68 % CL one.

Table 3.8 Relative systematic uncertainties from each source, along with the total relative systematic and statistical uncertainties, (in %), on the ADD signal yield

n	PDF	ISR/FSR	Fact. Ren.	JES/$E_{\mathrm{T}}^{\mathrm{miss}}$	Lumi.	Trig.	Tot. Syst.	MC stat.
Signal region 1								
2	4.0	6.0	24	4.1	3.9	1	25	0.79
3	6.8	7.6	25	3.1	3.9	1	28	0.73
4	9.7	6.9	26	2.4	3.9	1	29	0.72
5	12	5.4	26	4.9	3.9	1	29	0.71
6	14	4.8	25	6.6	3.9	1	30	0.72
Signal region 2								
2	4.3	2.7	30	7.4	3.9	0	31	1.5
3	6.8	7.9	25	5.9	3.9	0	28	1.2
4	9.5	3.3	26	4.6	3.9	0	28	1.2
5	12	5.3	30	4.5	3.9	0	33	1.2
6	14	3.2	27	4.0	3.9	0	31	1.1
Signal region 3								
2	4.8	6.7	30	9.8	3.9	0	33	2.8
3	6.9	7.1	31	8.0	3.9	0	34	2.2
4	9.4	2.8	24	7.7	3.9	0	27	2.0
5	12	6.2	36	6.9	3.9	0	39	1.9
6	14	2.8	34	5.6	3.9	0	36	1.8
Signal region 4								
2	5.6	8.7	30	12	3.9	0	34	5.2
3	7.2	12	31	9.4	3.9	0	36	4.0
4	9.5	6.0	24	8.7	3.9	0	28	3.6
5	12	14.0	36	10.0	3.9	0	42	3.3
6	13	9.6	34	9.5	3.9	0	40	3.1

Limits based on SR1, SR2 and SR3 are typically 35, 15 and 5 % worse than SR4 respectively.

Figure 3.30 shows the LO visible cross section of the ADD signal corresponding to SR4 as a function of the scale M_D, for $n = 2, 4$ and 6. The signal theoretical uncertainties are shown as colored bands. The model-independent 95 % CL expected and observed limits on σ_{vis} are shown as horizontal lines. Expected limits are recomputed varying the BG prediction by $\pm 1\sigma$ of the statistical and systematic uncertainties on the BG itself. The effect of this variation is shown by the grey band around the expected limits.

Expected and observed limits on M_D are calculated taking into account all the uncertainties on the BG predictions, as well as the experimental uncertainties on the signal, but not the signal theoretical uncertainties. Correlations between the experi-

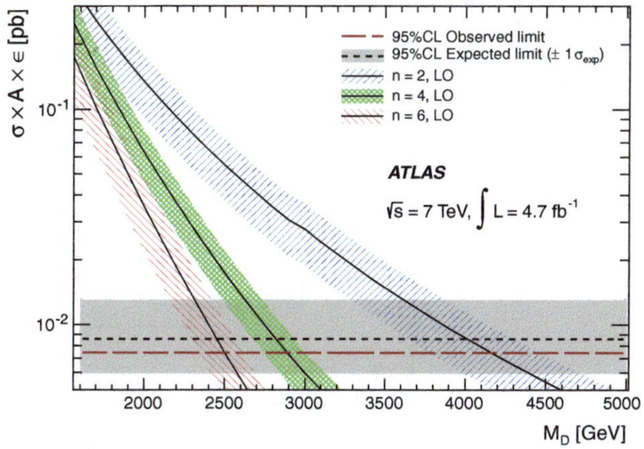

Fig. 3.30 Visible cross section as a function of the scale M_D, compared with the expected and observed limits. The theoretical uncertainties are shown as *colored bands*, while the experimental uncertainties are shown as the *grey band* around the expect limit

Table 3.9 Final observed limits on the ADD model based on the results of SR4

n	M_D (TeV)		R (pm)		Truncation effect	
	LO	NLO	LO	NLO	LO (%)	NLO (%)
2	4.17	4.37	2.8×10^7	2.5×10^7	0.02	0.01
3	3.32	3.45	4.8×10^2	4.5×10^2	1.9	1.3
4	2.89	2.97	2.0	1.9	11.8	9.9
5	2.66	2.71	7.1×10^{-2}	7.0×10^{-2}	29.5	27.2
6	2.51	2.53	0.8×10^{-2}	0.8×10^{-2}	49.1	47.9

The limits are split into those retrieved with LO and NLO cross section. The effect of truncating the cross section calculation for the phase space in which $\hat{s} > M_D^2$ is shown in the last columns

mental uncertainties on signal and BG are taken into account. Exclusion limits are calculated considering both LO and NLO pQCD cross sections.

Results based on 4.7 fb^{-1} of 7 TeV exclude at 95 % CL values of M_D lower than 4.37 TeV for $n = 2$, and lower than 2.53 TeV for $n = 6$ (using NLO signal cross sections). The results using both LO and NLO pQCD signal predictions are presented in Table 3.9 and are shown in Figs. 3.31 and 3.32, as a function of n. For comparison, the figures show also the previous ATLAS limits based on 33 pb^{-1}. Expected limits are recomputed varying the BG and signal yields by $\pm 1\sigma$ of their experimental uncertainty, and the effect is represented by the grey band around the expected limit. In order to show the effect of the theoretical uncertainties on the exclusion limits, the signal yields are varied by $\pm 1\sigma_{THEO}$ and the observed limits are recomputed (shown in the figures by the dashed lines around the nominal observed limits). For example, for $n = 2$ the 95 % CL lower limit on M_D are 4 % lower when the signal yield is varied by $-1\sigma_{THEO}$ (conservative bound).

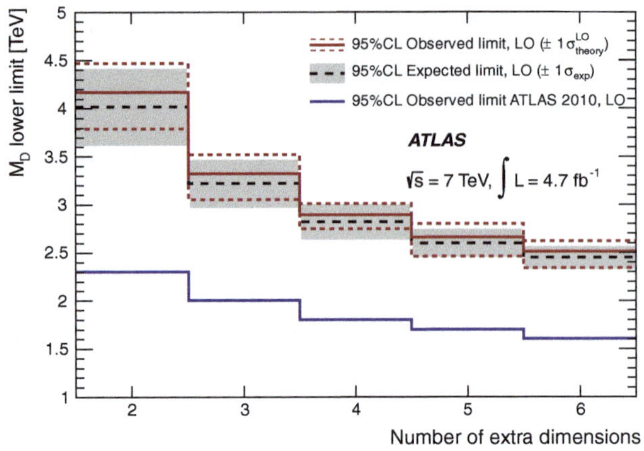

Fig. 3.31 Limits on M_D as a function of the number of extra dimension n. The limits are calculated from LO cross sections

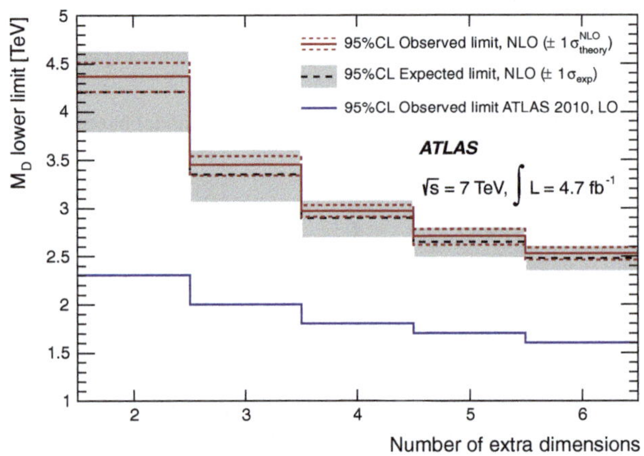

Fig. 3.32 The plot on the *left-hand side* shows the limits on M_D as a function of the number of extra dimension n. The limits are calculated from NLO cross sections

Finally, Fig. 3.33 presents a comparison between the observed limits computed from LO and NLO cross sections. The exclusion bound for $n = 2$ ($n = 6$) from the NLO cross section is 5 % (1 %) higher compared to LO one, and the effect of the theoretical uncertainty on the limits is reduced by ~50 % independently of n.

The constraints on M_D can be translated into limits on the extra-dimension radius R. Considering the same radius for all extra-dimensions, R is related to M_D as

$$R = \frac{1}{M_D} \cdot [\frac{M_{pl}}{\sqrt{8\pi} M_D}]^{\frac{2}{n}}, \tag{3.2}$$

where M_{pl} is the Planck mass. For $n = 2$ ($n = 6$) and using NLO cross sections, radius larger than 25 μm (8 fm) are excluded at 95 % CL. Other exclusion limits on R are reported in Table 3.9.

The effective field theory employed to describe the ADD signal is expected to be valid only if the scales involved in the hard interaction are significantly smaller

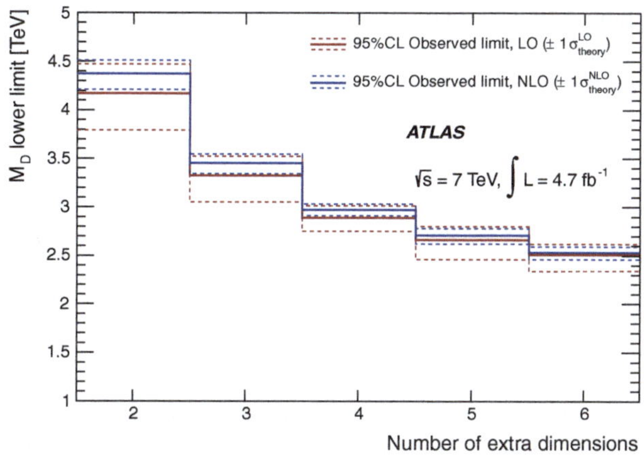

Fig. 3.33 The plot on the *right-hand side* shows a comparison between the observed limits retrieved from LO and NLO cross sections

than M_D. In order to approximately quantify the impact on the limits from the phase space where the effective field theory is suspected to be no longer valid, we recalculate the cross sections rejecting the events with $\hat{s} > M_D^2$, where $\sqrt{\hat{s}}$ is the center-of-mass energy of the hard interaction. The relative difference between "truncated" and complete cross sections gives an estimate of the reliability of the effective field theory. This difference increases from SR1 to SR4 because of the increasing thresholds in E_T^{miss} and leading jet p_T. For SR4 and $n = 2$, the difference between the truncated and the un-truncated cross sections is negligible, while for $n = 3, 4, 5$, and 6 differences of the order of 2, 12, 30, and 50 % are found (see Table 3.9). This indicates that the exclusion limits reported have a large sensitivity to the implementation of the effective field theory.

Here, the final limits on M_D are calculated considering the full cross sections. Other alternative approaches were used in the previous ATLAS publications. For example, in the analysis of the first 33 pb^{-1} [2], that is described in Appendix C, the limits where quoted only for $n = 2$, 3, and 4, where the effect of the ultra-violet theory is relatively small. Instead, for the analysis of the first 1 fb^{-1} [3], described in appendix D, the limits where retrieved from a SR with moderate thresholds in E_T^{miss} and leading jet p_T ($E_T^{miss} > 220$ GeV and $p_T > 250$ GeV), and therefore reducing the dependency of the results from the unknown ultra-violet limit of the theory. Another alternative approach is to suppress the cross section in the phase space with $\hat{s} > M_D^2$. The CMS collaboration has adopted this strategy, using a suppression factor $M_D^2/\hat{s}$.

3.7.2 Limits on WIMP Pair Production

The results based on 4.7 fb^{-1} of 7 TeV data are translated into exclusion limits on WIMP pair production, in the framework of an effective lagrangian with contact interactions between WIMPs and SM particles. A detailed description of the

Table 3.10 Effective operators for Dirac fermion WIMPs coupling to SM quarks or gluons

Name	Coupling	Operator
D1	Scalar	$\frac{m_q}{(M^*)^3}\bar{\chi}\chi\bar{q}q$
D5	Vector	$\frac{1}{(M^*)^2}\bar{\chi}\gamma^{\mu}\chi\bar{q}\gamma_{\mu}q$
D8	Axial-vector	$\frac{1}{(M^*)^2}\bar{\chi}\gamma^{\mu}\gamma^5\chi\bar{q}\gamma_{\mu}\gamma^5 q$
D9	Tensor	$\frac{1}{(M^*)^2}\bar{\chi}\sigma^{\mu\nu}\chi\bar{q}\sigma_{\mu\nu}q$
D11	Scalar	$\frac{1}{(4M^*)^3}\bar{\chi}\chi\alpha_s(F_{\alpha\beta}F^{\alpha\beta})$

χ defines the Dirac-like WIMP field, while q and $F^{\alpha\beta}$ represent the quark and the gluon strength tensor

phenomenology of WIMP pair production at colliders, and the implementation of effective interactions was already given in Sect. 1.4.2. Table 3.10 reports the list of effective operators that are employed in this analysis to describe possible interactions between WIMPs (assumed to be Dirac-like particles) and quarks or gluons. The different operators will contribute to either spin-dependent or spin-independent WIMP-nucleon interactions, for which exclusion limits will be evaluated as a function of the WIMP mass m_χ. Results will be translated also into exclusion limits on the WIMP-WIMP annihilation rate (see Sect. 1.4.2).

Systematic uncertainties The systematic uncertainties have been treated similarly to those of the ADD limits. The experimental uncertainty on $E_{\mathrm{T}}^{\mathrm{miss}}$/JES and JER results in a variations of the signal yields between 1 and 20 % depending on the operator and the SR considered.[10] The uncertainties on the trigger efficiency and on the integrated luminosity introduce an uncertainty on the signal yield of 1 % (for SR1 only) and 3.8 % (for the four SRs), respectively.

The WIMP MC samples use the PDF set CTEQ6L1.[11] In order to assess the uncertainties on the choice of this PDF set, the 44 error sets associated to CTEQ6M are employed, and the relative difference respect to the nominal CTEQ6M set is adopted. The uncertainty on the signal yield varies between 3 % and 17 % depending on the operator.

For the uncertainty on the ISR/FSR modeling, the jet matching scale between MADGRAPH and PYTHIA (see Sect. 1.3) is varied by a factor two and one half. In addition the parameters that regulate the PS in PYTHIA have been varied in the range allowed by experimental data. This introduces an uncertainty on the signal yield that varies between 3 and 6 %, affecting only the acceptances.

The uncertainty on the factorization and renormalization scales is the dominant theoretical uncertainty. Variations of the scales by factors two and one-half, result in changes of 30 % on the signal yield, increasing with increasing $E_{\mathrm{T}}^{\mathrm{miss}}$. A summary of the uncertainties on the WIMP signal is presented in Table 3.11.

---

[10] Very low JES uncertainties come from the compensation of effects on the requirements on the leading and the veto on sub-leading jets.

[11] The use of CTEQL1 was conditioned by the underlying event tune employed.

Table 3.11 Summary of WIMP signal relative uncertainties, for the different operators and for the four SRs

Uncertainty	Operator	SR1	SR2	SR3	SR4
ISR/FSR	All	4.4	5.2	6.3	6.3
Jet matching scale	All	2.8	2.8	2.8	5.1
Fact. Renor. Scales	All	31.5	31.5	31.5	31.5
PDF	D1	17.6	17.6	17.6	17.6
	D5	4.9	4.9	4.9	4.9
	D9	3.6	3.6	3.6	3.6
	D11	16.4	16.4	16.4	16.4
JES/JER/$E_{\mathrm{T}}^{\mathrm{miss}}$	D1	6.4	8.8	13.7	18.8
	D5	5.2	7.9	10.3	17.2
	D9	3.3	5.1	6.5	12.2
	D11	0.8	2.3	5.6	10.0
Luminosity	All	3.9	3.9	3.9	3.9
Trigger	All	1	0	0	0

Limits on the suppression scale M* The results of the analysis are translated into 90 % CL and 95 % CL exclusion limits on the suppression scale M^* (see Sect. 1.4.2) as a function of the WIMP mass m_χ. As for the ADD signal, the expected and observed limits are computed taking into account the uncertainties on the BG predictions and the experimental and the statistical uncertainties on the signal. The effect of the signal theoretical uncertainties is evaluated recomputing the observed limits varying the signal yields by $\pm 1\ \sigma_{THEO}$ (theoretical systematic uncertainty on the signal). The results from SR3 or SR4 have been used for the limit setting depending on the best expected limit. For operator D1, D5 and D8, the results from SR3 are used, while for D9 and D11 results from SR4 are employed. Limits from SR1 and SR2 are typically 40 and 15 % worse.

The exclusion limits are shown in Table 3.12 and Fig. 3.34 for the different operators and as a function m_χ. In the case of $m_\chi = 5$ GeV values of M^* lower than 30 GeV (687 GeV) for operator D1 (D5) are excluded at 90 % CL, while for $m_\chi = 700$ GeV the same limits are at 14 GeV (416 GeV). As it can be seen in the figures, for values of m_χ lower than 10 GeV the limits do not essentially depend on m_χ.

The "thermal relic" green line in Fig. 3.34 corresponds to the values of M^* and m_χ for which WIMPs would annihilate to SM particles resulting exactly in the thermal relic density observed with WMAP [38] (see Sect. 1.4.2).

As discussed in Sect. 1.4.2, the effective field theory is valid if the mediators of the interaction are not produced directly, and if the couplings between WIMP and SM particles, and mediators are not too strong so that the interaction can be described in perturbation theory. From these considerations we infer that the effective field theory

Table 3.12 90 % (95 %) CL observed lower limits on the suppression scale M^* for different values of the WIMP mass m_χ

	Limits on M^* (GeV)				
m_χ (GeV)	D1	D5	D8	D9	D11
1	30 (29)	687 (658)	687 (658)	1353 (1284)	375 (361)
5	30 (29)	687 (658)	687 (658)	1353 (1284)	375 (361)
10	30 (29)	687 (658)	687 (658)	1353 (1284)	375 (361)
50	30 (29)	682 (653)	666 (638)	1338 (1269)	370 (357)
100	29 (28)	681 (653)	650 (623)	1310 (1243)	360 (347)
200	27 (26)	658 (631)	595 (570)	1202 (1140)	357 (344)
400	21 (20)	571 (547)	475 (455)	943 (893)	324 (312)
700	14 (14)	416 (398)	311 (298)	629 (596)	250 (241)
1000	9 (9)	281 (269)	196 (188)	406 (384)	185 (178)
1300	6 (6)	173 (165)	110 (106)	240 (227)	128 (123)

All values are expressed in GeV

is not valid in the region of the parameter space $m_\chi \geq 2\pi M^*$. This region is shown with a grey area in the bottom right corner of the figures.

Limits on WIMP-nucleon cross section Constraints on M^* are translated into limits on WIMP-nucleon scattering cross section adopting the procedure explained in Sect. 1.4.2 and in Ref. [22]. These limits are compared with those from direct detection experiments.

Results from direct detection and from WIMP-pair production at colliders are extracted from very different regimes, and in the latter case the prediction might suffer from the inadequacy of the effective theory approach. The typical transferred momentum $\sqrt{Q^2}$ considered in a direct detection experiment is at the order of a keV. In this regime the propagator of a particle with mass $m \gg 1$ KeV that mediate the interaction cannot be resolved, making a Fermi-like point interaction suitable. On the contrary, at LHC we probe energy scales up to few TeV. It is important to remark also that the ATLAS limits are done considering flavour-universal coupling between WIMPs and the four lightest quarks, and that the interaction is assumed to happen through only one of the considered operators.

Figures 3.35 and 3.36 show the 90 % upper limits on the WIMP-nucleon cross section as a function of m_χ, for spin-independent and spin-dependent operators, respectively. As for the ADD signal, the observed limits are calculated taking into account all but the theoretical uncertainty on the signal. In these plots the effect of the theoretical uncertainties is shown by recomputing the limits with the signal yield varied by $-1\sigma_{THEO}$ (conservative bound). For 5 GeV$< m_\chi <$ 100 GeV and for operators D1 (scalar) and D5 (vector) the ATLAS results exclude at 90 % CL cross sections larger than 10^{-38} and $5 \times 10^{-39}\,\text{cm}^{-2}$. Due to kinematic constraints, the ATLAS limits are less powerful for $m_\chi >$ 500 GeV. Exclusion limits from operator

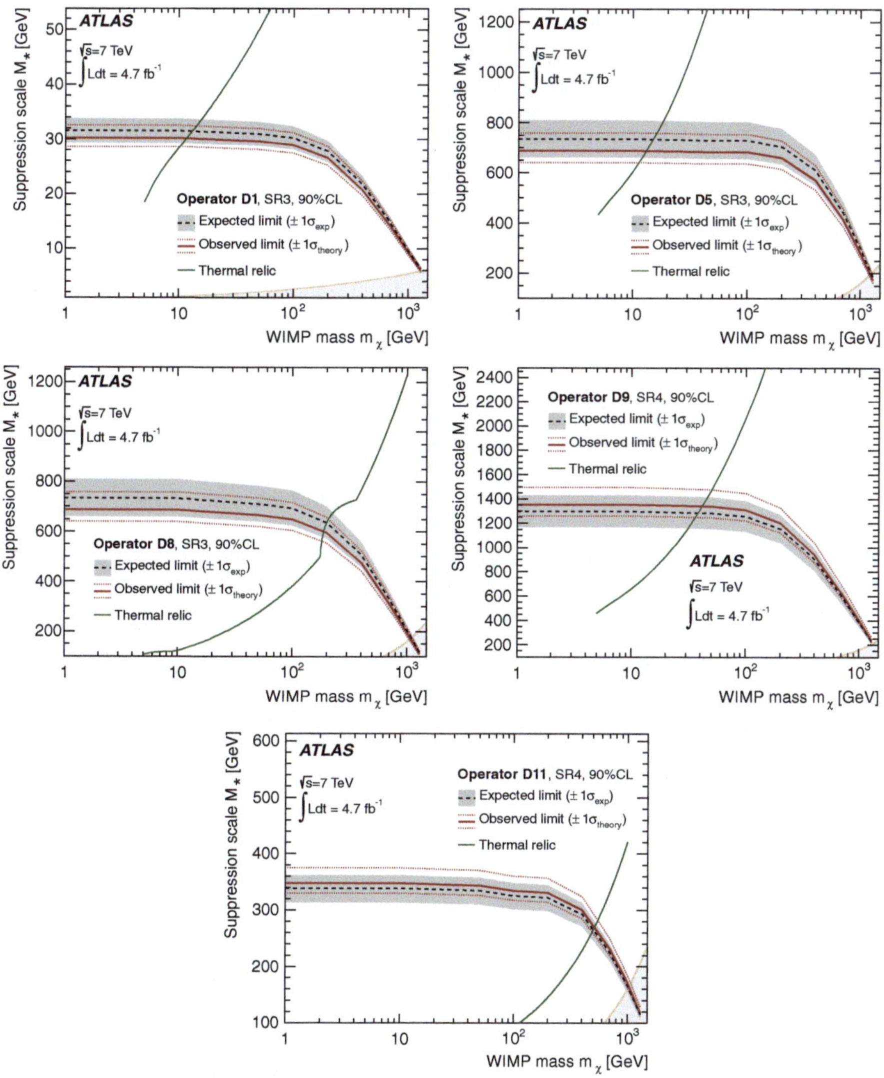

Fig. 3.34 90 % CL expected and observed lower limits on M^* as a function of the WIMP mass m_χ. The region where the effective field theory breaks down is shown as *filled gray area* in the *bottom right corner* of the figures. The *green lines* show the M^* values at which WIMPs would result in the required relic abundance

D11 (scalar) are particularly enhanced by the fact that the WIMP production is driven by gluon interactions.

For the spin-independent interactions (Fig. 3.35) the ATLAS limits are compared with results from XENON100 [39], CDMSII [40], CoGeNT [41], CDF [42] and CMS [43]. The collider results are particularly relevant for low WIMP masses

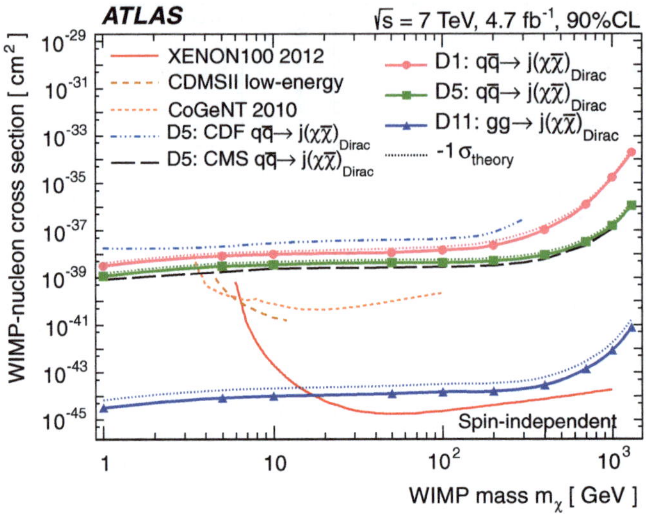

Fig. 3.35 90 % CL WIMP-nucleon cross section upper limits as a function of the WIMP mass m_{χ}. The ATLAS results for different spin-independent operators are compared with the results from XENON100 [39], CDMSII [40], CoGeNT [41], CDF [42] and CMS [43]

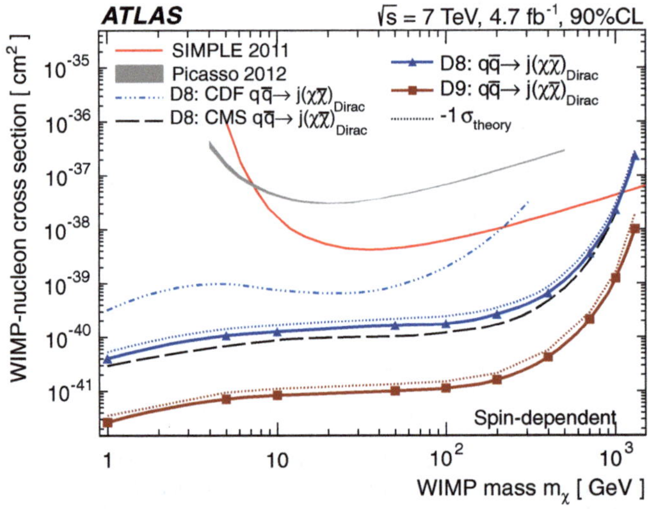

Fig. 3.36 90 % CL WIMP-nucleon cross section upper limits as a function of the WIMP mass m_{χ}. The ATLAS results for different spin-dependent operators are compared with the results from SIMPLE [44], Picasso [45], CDF [42] and CMS [43]

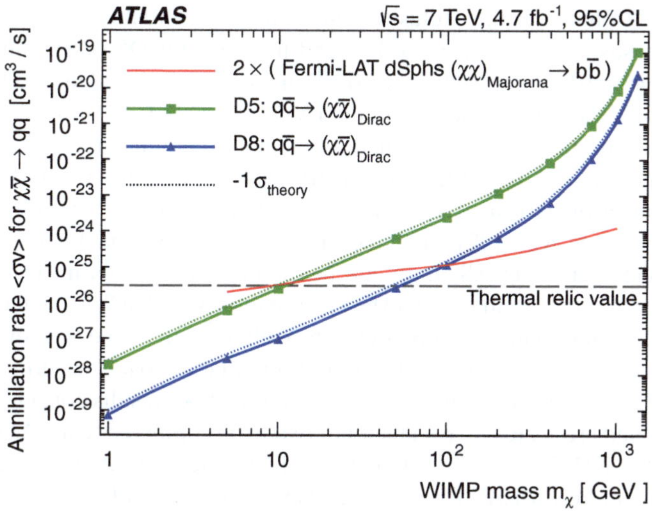

Fig. 3.37 90 % CL observed limits on $\langle \sigma v \rangle$ as a function of the WIMP mass m_χ. The quantity $\langle \sigma v \rangle$ is defined as in Ref. [48]. The ATLAS results for the D5 and D8 operators are compared with the high-energy gamma-ray limits from observations of Galactic satellite galaxies with the Fermi-LAT experiment [46]. The latter limits are inferred on Majorana-like WIMPs, and they have been scaled up by a factor of two to be compared with the ATLAS result on Dirac-like WIMPs. The $\langle \sigma v \rangle$ value for which WIMPs to make up the relic abundance set by the WMAP measurement is indicated by the *horizontal dashed line*

($m_\chi < 5$ GeV), where the direct detection experiments are less sensitive due to kinematic suppression. For $m_\chi > 10$ GeV and in the case of the operator D11 the ATLAS limits are the most performant up to $m_\chi = 20$ GeV and are competitive with direct detection limits up to $m_\chi \sim 1$ TeV. As discussed in Sect. 1.4.2, the results by DAMA, CoGeNT and other experiments can be interpreted as a signal from a WIMP-nucleon interaction with cross section of $\sim 10^{-40}$ cm^{-2} with a $m_\chi = 5$–10 GeV. This value is beyond the current exclusion of ATLAS.

For the spin-dependent operators (Fig. 3.36) the ATLAS limits are compared with the results from SIMPLE [44], Picasso [45], CDF [42] and CMS [43]. The results from colliders are the most relevant for the exclusion for all WIMP masses up to 1 TeV.

Limits on WIMP-WIMP annihilation rate Lower bounds on M^* are translated into WIMP-WIMP annihilation rate and compared to indirect detection results under some assumptions. The annihilation rate $\langle \sigma v \rangle$ is defined as the product of the annihilation cross section for two WIMP particles and their relative velocity, and is averaged over the WIMP velocity spectrum (see Sect. 1.4.2). The conversion from the M^* limits has been made with the formulas 1.22 and 1.23. Here, WIMPs are assumed to annihilate exclusively to the four light quarks with the same coupling for all flavors, and only one interaction at the time.

Figure 3.37 illustrates the exclusion limits on the annihilation rate for D5 and D8 operators versus m_χ. For $m_\chi = 5\,\text{GeV}$ and the operator D5 (D8), values of $\langle\sigma v\rangle$ higher than $6 \times 10^{-27}\,\text{cm}^2\,\text{s}^{-1}$ ($3 \times 10^{-28}\,\text{cm}^2\,\text{s}^{-1}$) are excluded at 90 % CL. For $m_\chi = 700\,\text{GeV}$ the corresponding limit is $10^{-21}\,\text{cm}^2\,\text{s}^{-1}$ ($10^{-22}\,\text{cm}^2\,\text{s}^{-1}$) for the D5 (D8) operator. In the figure, the dashed horizontal green line labeled "thermal relic value", shows the $\langle\sigma v\rangle$ for which the WIMP-WIMP annihilation would result in the right abundance of dark matter in the early universe compatible with the WMAP measurements (see Ref. [22]). For the operator D5 (D8) and for $m_\chi < 10\,\text{GeV}$ ($m_\chi < 50\,\text{GeV}$) the ATLAS limits are below this value. This means that WIMP masses lower than these values would be incompatible with the WMAP measurement, or that WIMPs annihilate to SM particles through more than one type of interaction. Figure 3.37 reports also results on the WIMP annihilation to $b\bar{b}$ by Fermi Large Area Telescope (Fermi-LAT) experiment [46]. The Fermi-LAT results consider Majorana-like WIMPs and are scaled up by a factor of two to be compared to the ATLAS results on Dirac-like WIMPs (see Eq. 34 in Ref. [47]). The ATLAS limits are stronger than those from Fermi-LAT for $m_\chi < 10$ GeV in the case of the D5 operator, and $m_\chi < 100$ GeV in the D8 scenario.

References

1. G. Aad et al., Search for dark matter candidates and large extra dimensions in events with a jet and missing transverse momentum with the ATLAS detector. JHEP **04** (2013)
2. G. Aad et al., Search for new phenomena with the monojet and missing transverse momentum signature using the ATLAS detector in $\sqrt{s} = 7\,\text{TeV}$ proton-proton collisions. Phys. Lett. B **705**, 294–312 (2011). doi:10.1016/j.physletb.2011.10.006
3. Search for new phenomena in monojet plus missing transverse momentum final states using 1 fb^{-1} of pp collisions at $\sqrt{s} = 7\,\text{TeV}$ with the atlas detector (ATLAS-CONF-2011-096), July 2011
4. D. Casadei et al., The implementation of the atlas missing et triggers for the initial lhc operation (ATL-DAQ-PUB-2011-00) (2011)
5. S. Agostinelli et al., GEANT4: A simulation toolkit. Nucl. Instrum. Meth. **A506**, 250–303 (2003). doi:10.1016/S0168-9002(03)01368-8
6. ATLAS Collaboration, The ATLAS simulation infrastructure. Eur. Phys. J. **C70**, 823–874 (2010)
7. M. Mangano et al., Alpgen, a generator for hard multiparton processes in hadronic collisions. JHEP **07**, 001 (2003)
8. G. Corcella et al., HERWIG 6: An event generator for hadron emission reactions with interfering gluons (including supersymmetric processes). JHEP **01**, 010 (2001)
9. G. Corcella et al., HERWIG 6.5 release note (2002). hep-ph/0210213
10. J. Forshaw, J. Butterworth, M. Seymour, Multiparton interactions in photoproduction at hera. Z. Phys. **C72**, 637–646 (1996)
11. J. Huston, H. Lai, P.M. Nadolsky, J. Pumplin, D. Stump et al., New generation of parton distributions with uncertainties from global QCD analysis. JHEP **012**, 0207 (2002)
12. F. Krauss, M. Schonherr, S. Schumann, T. Gleisberg, S. Hoeche et al., Event generation with SHERPA 1.1. JHEP **007**, 0902 (2009)
13. R. Gavin, Y. Li, F. Petriello, S. Quackenbush, FEWZ 2.0: A code for hadronic Z production at next-to-next-to-leading order (2010)
14. S. Frixione, B.R. Webber, The MC@NLO 3.2 event generator (2006)

15. Pavel M. Nadolsky et al., Implications of CTEQ global analysis for collider observables. Phys. Rev. D **78**, 013004 (2008). doi:10.1103/PhysRevD.78.013004
16. S. Mrenna, T. Sjostrand, P. Skands, PYTHIA 6.4 physics and manual. JHEP **05**, 026 (2006)
17. J.M. Campbell, R.K. Ellis, Phys. Rev. D **65**, 113007 (2002)
18. A.D. Martin, W.J. Stirling, R.S. Thorne, G. Watt, Parton distributions for the LHC. Eur. Phys. J. **C63**, 189–285 (2009). doi:10.1140/epjc/s10052-009-1072-5
19. G.F. Giudice, R. Rattazzi, J.D. Wells, Quantum gravity and extra dimensions at high-energy colliders (2000)
20. Q. Li, S. Karg, M. Kramer, D. Zeppenfeld, NLO QCD corrections to graviton production at hadron colliders. Phys. Rev. D **81**, 094036 (2010)
21. F. Maltoni, O. Mattelaer, J. Alwall, M. Herquet, T. Stelzer, Madgraph 5: Going beyond. JHEP **1106**, 128 (2011)
22. J. Goodman, M. Ibe, A. Rajaraman, W. Shepherd, T.M.P. Tait et al., Constraints on dark matter from colliders. Phys. Rev. D **82**, 116010 (2010). doi:10.1103/PhysRevD.82.116010
23. M. Cacciari, G.P. Salam, G. Soyez, The anti-k_t jet clustering algorithm. JHEP **04**, 063 (2008). doi:10.1088/1126-6708/2008/04/063
24. G. Aad et al., Jet energy measurement with the ATLAS detector in proton-proton collisions at $\sqrt{s} = 7$ TeV (2011)
25. T. Barillari et al., ATL-LARG-PUB-2009-001
26. G. Aad et al., Performance of missing transverse momentum reconstruction in proton-proton collisions at 7 TeV with ATLAS. Eur. Phys. J. **C72**, 1844 (2012). doi:10.1140/epjc/s10052-011-1844-6
27. G. Aad et al., Electron performance measurements with the ATLAS detector using the 2010 LHC proton-proton collision data. Eur. Phys. J. **C72**, 1909 (2012) (Long author list—awaiting processing)
28. G. Aad et al., Measurement of the W $\rightarrow$ with the ATLAS detector. JHEP **1012**, 060 (2010). doi:10.1007/JHEP12(2010).060 (38 pages plus author list (57 pages total), 16 figures, 15 tables)
29. The ATLAS Collaboration, Data-quality requirements and event cleaning for jets and missing energy reconstruction with the atlas detector in proton-proton collisions at a center-of-mass energy of $\sqrt{s} = 7$ TeV (ATLAS-CONF-2010-038), June 2010
30. ATLAS Collaboration, Search for mono-jet plus missing transverse energy in pp with the ATLAS detector (ATL-COM-PHYS-2012-343), May 2012
31. The ATLAS Collaboration, Measurement of the cross section for jets produced in association with z bosons (ATL-COM-PHYS-2010-884), December 2010
32. The ATLAS Collaboration, Measurement of the production cross for w-bosons in association with jets in pp collisions at $\sqrt{s} = 7$ TeV with the atlas detector (ATL-COM-PHYS-2010-883), December 2010
33. ATLAS Internal, Single boson and diboson production cross sections in pp collisions at $\sqrt{s} =$ 7 TeV (ATL-COM-PHYS-2010-695)
34. ATLAS Collaboration, Non-collision backgrounds as measured by the ATLAS detector during the 2010 proton-proton run (ATLAS-CONF-2011-137), August 2011
35. A.L. Read, Presentation of search results: The CLs technique. J. Phys. G **G28**, 2693–2704 (2002)
36. N. Arkani-Hamed, S. Dimopoulos, G.R. Dvali, The hierarchy problem and new dimensions at a millimeter. Phys. Lett. B **429**, 263 (1998)
37. J. Pumplin, D. Stump, R. Brock, D. Casey, J. Huston et al., Uncertainties of predictions from parton distribution functions. 2. The Hessian method. Phys. Rev. D **65**, 014013 (2001). doi:10.1103/PhysRevD.65.014013
38. E. Komatsu et al., Seven-year Wilkinson microwave anisotropy probe (WMAP) observations: Cosmological interpretation. Astrophys. J. Suppl. **192**, 18 (2011). doi:10.1088/0067-0049/192/2/18
39. XENON100 Collaboration, Dark matter results from 225 live days of data
40. CDMS Collaboration, Results from a low-energy analysis of the cdms ii germanium data. Phys. Rev. Lett. **106**, 131302 (2011)

41. C. Aalseth et al., CoGeNT Collaboration. Results from a search for light-mass dark matter with a p-type point contact germanium detector. Phys. Rev. Lett. **106**, 131301 (2011)
42. T. Aaltonen et al., CDF Collaboration, Search for dark matter in events with one jet and missing transverse energy in pp collisions at s = 1.96 TeV. Phys. Rev. Lett. **108**(21), 211804 (2012)
43. CMS Collaboration, Search for dark matter and large extra dimensions in monojet events in pp collisions at s = 7 TeV
44. T. Morlat, A. Fernandes, A. Ramos, M. Felizardo, T. Girard et al., Final analysis and results of the phase ii simple dark matter search. Phys. Rev. Lett. **108**, 201302 (2012)
45. PICASSO Collaboration, Constraints on low-mass wimp interactions on 19f from picasso. Phys.Lett. B, **711**, 153–161 (2012)
46. M. Ackermann et al., Constraining dark matter models from a combined analysis of Milky Way satellites with the Fermi Large Area Telescope. Phys. Rev. Lett. **107**, 241302 (2011) (6 pages, 2 figures/ Contact authors: Johann Cohen-Tanugi, Jan Conrad, and Maja Llena Garde)
47. M. Cirelli et al., Pppc 4 dm id: A poor particle physicist cookbook for dark matter indirect detection
48. P.J. Fox, R. Harnik, J. Kopp, Y. Tsai, Missing energy signatures of dark matter at the LHC. Phys. Rev. D **85**, 056011 (2012). doi:10.1103/PhysRevD.85.056011 (22 pages, 10 figures)

Chapter 4
Search for New Phenomena in the Mono-jet Final State at $\sqrt{s} = 8\,\text{TeV}$

This chapter describes the mono-jet analysis performed with 10.5 fb^{-1} of p-p collisions at $\sqrt{s} = 8\,\text{TeV}$ collected in 2012. This analysis was presented for the first time in the HCP2012 conference and is documented in reference [1]. It follows other publications that use $\sqrt{s} = 7\,\text{TeV}$ data collected in 2010 and 2011 [2–4]. The analysis strategy is almost unchanged compared to that described in Chap. 3. Therefore, in this chapter we will mainly point-out the differences with respect to the analysis of the 7 TeV dataset, referring to the previous chapter for the rest.

This chapter is organized as follows. Sections 4.1 and 4.2 present the various aspects of the ATLAS dataset and simulated MC samples used for the analysis. The definition of the physics objects and the event selection criteria are detailed in Sect. 4.3. The background estimation is discussed in Sect. 4.4, and results are presented in Sect. 4.5. Section 4.6 focusses the interpretation of the results in the context of the ADD extra-dimension model, WIMP pair production, and gravitino squark/gluino associated production in the GMSB scenario.

4.1 Data Sample

This analysis makes use of nearly half of the 8 TeV dataset recorded by ATLAS in 2012. After applying basic data quality requirements on the data-taking conditions the remaining dataset consists of 10.5 fb^{-1}. The maximum instantaneous luminosity increased from 2.7×10^{30} to $7.6 \times 10^{33}\,\text{cm}^{-2}\,\text{s}^{-1}$ with time. This translated into an increase in the mean number of collisions per bunch crossing from 5 to 36.

Trigger selection Events are collected with the lowest unprescaled $E_{\text{T}}^{\text{miss}}$ trigger, called "EF_xe80_tclcw", that uses only calorimeter-based quantities with no corrections for identified muons. This trigger item has two advantages compared to the one used for the 7 TeV analysis. First, the level-2 algorithm provides a better $E_{\text{T}}^{\text{miss}}$ reconstruction using the cell energy reconstruction, instead of the simplified version

V. Rossetti, *Search for Exotic Mono-jet Events*, Springer Theses,
DOI 10.1007/978-3-319-22225-7_4

used at level-1.[1] Second, the Event Filter uses the LCW E_T^{miss} calibration, as it is done for the offline E_T^{miss} (see Sect. 2.3.4). These two improvements provide a 10 to 15 % increase in trigger efficiency with respect to the 7 TeV period. The E_T^{miss} thresholds at the three trigger levels are 60, 65, and 80 GeV at the level-1, level-2, and Event Filter, respectively. These thresholds are higher compared to the 7 TeV trigger configuration, to handle the increase of the instantaneous luminosity. This results in a loss of efficiency for the mono-jet analysis that is partially compensated by the improvements in the trigger algorithms, discussed above.

4.2 Monte Carlo Simulated Samples

The MC production at 8 TeV follows closely what was done for the 7 TeV samples (see Sect. 3.2). In order to account for pile-up effects all samples are generated with minimum bias interactions overlaid with the hard scattering event. The MC samples at 8 TeV are produced with a distribution of μ (average number of interactions per bunch crossing) that is not the same as in the data. The μ distribution of the MC is then corrected to exactly match that measured in data.

A set of $W(\rightarrow \ell\nu)$+jet, $Z/\gamma^*(\rightarrow \ell\ell)$+jet and $Z(\rightarrow \nu\nu)$+jet MC samples is produced with ALPGEN interfaced with HERWIG for the parton shower and hadronization, and JIMMY to simulate the underlying event. A separate set of Z/W+jets samples produced with SHERPA is used in the analysis to assess systematic uncertainties. These samples are produced with the MENLOPS technique (see Sect. 1.3). These ALPGEN and SHERPA samples are then normalized to inclusive Drell-Yan and W cross sections calculated at NNLO in perturbative QCD, as determined with the FEWZ program. Other SM processes are simulated with MC@NLO (single top and $t\bar{t}$), PYTHIA (QCD multi-jet), and HERWIG (di-bosons).

MC samples for ADD signals (see Sect. 1.4.1) are generated with PYTHIA. The production of WIMP pairs plus jets (see Sect. 1.4.2) is simulated with MADGRAPH5 with LO matrix elements [5] for up to two outgoing partons from ISR, and PYTHIA is used for the parton shower. The 8 TeV analysis focuses on the effective operators D5 and D11 (described in Sect. 1.4.2). For each operator, the MC samples are produced with WIMP masses of 80 GeV, 400 GeV and 1 TeV.

The results from the 8 TeV data are also interpreted in the context of a GMSB scenario (see Sect. 1.4.3). The MC samples for gravitino production in association with a squark or a gluino, $pp \rightarrow \tilde{G}\tilde{q} + X$ and $pp \rightarrow \tilde{G}\tilde{g} + X$, are generated with MADGRAPH using the PDF set CTEQ6L1, and interfaced with PYTHIA for the parton shower. The renormalization and factorization scales are set to the average

[1]The level-1 energy reconstruction is based on an analog sum of the calorimeter signals of several cells, and is less precise than the cell-by-cell energy reconstruction.

mass of the particles produced in the hard interaction, $(m_{\tilde{G}} + m_{\tilde{q},\tilde{g}})/2 \simeq m_{\tilde{q},\tilde{g}}/2$, where $m_{\tilde{G}}$, $m_{\tilde{q}}$, and $m_{\tilde{g}}$ are the masses of the gravitino, squark, and gluino, respectively. MC samples are produced with $m_{\tilde{G}}$ ranging between 10^{-3} and 10^{-5} eV, and $m_{\tilde{q}}$ and $m_{\tilde{g}}$ between 50 GeV and 2.6 TeV.

4.3 Object Definition and Event Selection

The physics object definition and selection criteria used in this analysis are very close to that of the 7 TeV analysis.

Jet candidates are selected with $p_{\mathrm{T}} > 30$ GeV and $|\eta| < 4.5$. Due to the high pile-up conditions in the 8 TeV data, the jet energy is corrected for multiple interaction effects before the final hadronic calibration is applied (see Sect. 2.3.3). The $E_{\mathrm{T}}^{\mathrm{miss}}$ is calculated from all calorimeter topo-clusters with $|\eta| < 4.5$, calibrated with the LCW scheme. Electrons are defined with $p_{\mathrm{T}} > 20$ GeV and $|\eta| < 2.47$, while muons are required to have $p_{\mathrm{T}} > 7$ GeV and $|\eta| < 2.5$. MC events are corrected to have the same electron and muon reconstruction efficiencies as measured in data. As for the 7 TeV analysis, these factors typically differ by the unity by less than 0.5 %.

Events are required to pass a set of cuts (described in Sect. 3.4) to reject events with fake $E_{\mathrm{T}}^{\mathrm{miss}}$ caused by non-collision BG and electronic noise. For the 8 TeV analysis, in order to cope with the increased non-collision BG, the leading jet is required to have $f_{max} < 0.8$, where f_{max} is the maximum fraction of jet energy deposited in one calorimeter layer.

As for the 7 TeV analysis, events are required to have at least one jet with $p_{\mathrm{T}} > 120$ GeV and $|\eta| < 2$, and no more than two jets with $p_{\mathrm{T}} > 30$ GeV and $|\eta| < 4.5$. Events with identified electrons or muons, or $\Delta\phi(jet2, E_{\mathrm{T}}^{\mathrm{miss}}) < 0.5$ are vetoed. Four signal regions (SRs) are then defined by requiring $E_{\mathrm{T}}^{\mathrm{miss}}$ and leading jet p_{T} above the thresholds 120, 220, 350 and 500 GeV.

The trigger selection is not fully efficient for the SR1 requirements. In order to determine the trigger efficiency we select an unbiased data sample collected with a different trigger (called "EF_mu18_medium") whose algorithms are based on the Muon Spectrometer. Events are required to pass the mono-jet selection of SR1, plus the identification of a muon with $p_{\mathrm{T}} > 25$ GeV. Figure 4.1 shows the efficiency of the "EF_xe80_tclcw" trigger in this event sample as a function of the offline reconstructed $E_{\mathrm{T}}^{\mathrm{miss}}$. The trigger is more than 97 % efficient for events in SR1, and data and MC agree within 1 %. To check the dependence of trigger efficiency with respect to pile-up, the study has been repeated separating the event sample in three categories based on the mean number of interactions per bunch crossing: $\mu \leq 10$, $10 < \mu \leq 20$, and $\mu > 20$. The trigger efficiency vary by less than 1 % between the categories, and the difference between data and MC is always below 1 %.

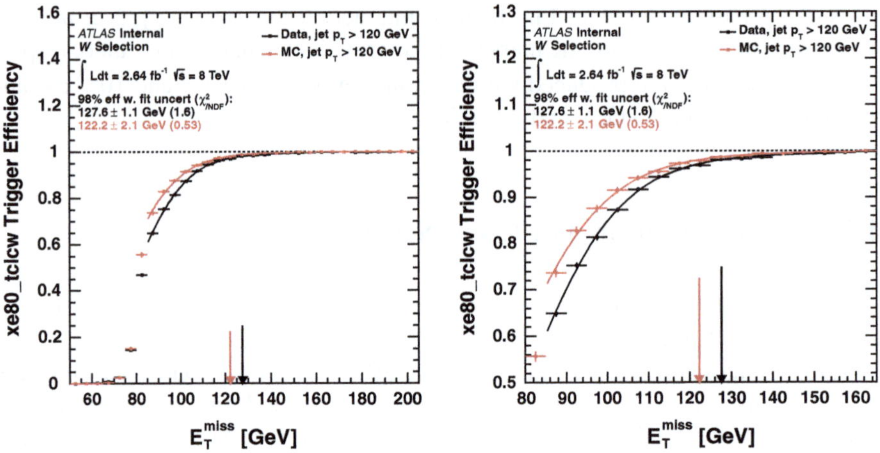

Fig. 4.1 Efficiency of the trigger used for this analysis (called "EF_xe80_tclcw") as a function of the offline $E_{\mathrm{T}}^{\mathrm{miss}}$. Data (*in black*) is compared to a $W(\rightarrow \mu\nu)$ Alpgen MC (*in red*). The two efficiency *curves* are fitted with an exponential function, and the *arrows* indicate the value of $E_{\mathrm{T}}^{\mathrm{miss}}$ for which the selection is 98 % efficient. On the *right*, a *zoomed view* of the *left plot* is shown. The figures are taken from Ref. [6]

4.4 Estimation of the Background Contributions

4.4.1 Z/W+jets Production

As for the analysis of the 7 TeV dataset, the Z/W+jet BG is estimated with a data-driven method that makes use of data events in control regions (CRs) and MC-based transfer factors (TFs) optimized to reduce the impact of the systematic uncertainties. For each SR we define corresponding CRs with the same selection but requiring the identification of muons or electrons. The inclusive electron CR requires at least one electron, while the $Z(\rightarrow \mu\mu)$ and $W(\rightarrow \mu\nu)$ CRs require respectively two muons with $76 < M_{\mu\mu} < 116$ GeV, and one muon with $40 < M_T < 100$ GeV. The estimated BG distribution in the SR are then built from the data distribution in the CRs applying bin-by-bin transfer factors (TFs) based on MC.

In this analysis, the $W(\rightarrow \mu\nu)$ CR is used for the estimation of all Z/W+jet processes, with the exception of $W(\rightarrow e\nu)$ and $Z(\rightarrow \tau\tau)$ for which the inclusive electron CR is employed. If the $Z(\rightarrow \mu\mu)$ control sample is used instead of the $W(\rightarrow \mu\nu)$ CR, consistent results are obtained. Similarly, consistent results are obtained for the $W(\rightarrow \tau\nu)$ BG contribution when the inclusive electron CR is used instead of the $W(\rightarrow \mu\nu)$ one.

Figures 4.2 and 4.3 show the reconstructed transverse mass of the W boson (before cutting on this variable), as well as the $E_{\mathrm{T}}^{\mathrm{miss}}$ and jet distributions measured in the inclusive electron and $W(\rightarrow \mu\nu)$ CRs. Figure 4.4 shows the invariant mass of the Z boson, and the $E_{\mathrm{T}}^{\mathrm{miss}}$ and jet distributions in the $Z(\rightarrow \mu\mu)$ CR. All the distributions

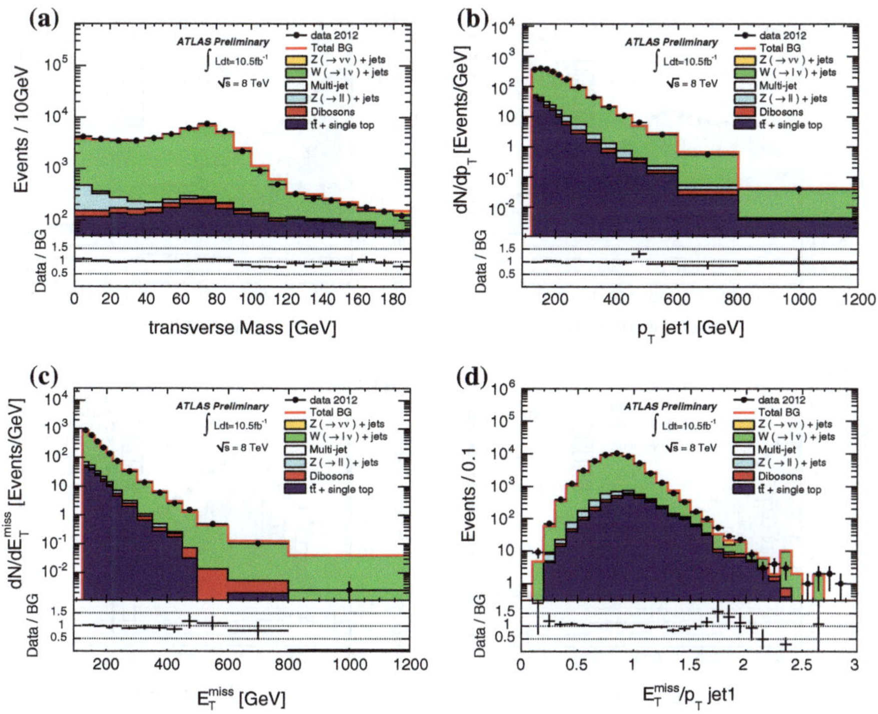

Fig. 4.2 Kinematic distributions of the events in the inclusive electron CR for the selection cuts of region 1 ($E_{\mathrm{T}}^{\mathrm{miss}}$, jet1 p_{T} >120 GeV)

show a reasonable agreement between data and MC. In all these figures, and only to put emphasis on the shape comparison, the Z/W+jet MC samples are normalized to the data. These global normalization factors are typically between 0.8 and 1.0 and are collected in Table 4.1.

As for the 7 TeV analysis, the Z/W+jets BG predictions include systematic uncertainties from: Z/W+jet MC modeling, jet and $E_{\mathrm{T}}^{\mathrm{miss}}$ related uncertainties, lepton identification, background subtraction in CRs, and trigger efficiency.

The uncertainties on the MC modeling of the Z/W+jet processes are estimated comparing results from ALPGEN and SHERPA MC samples. This comparison is carried out using two procedures. In the first one, we simply compare the two estimations on the total Z/W+jet BG. The differences range between $0.7 \pm 1.0\,\%$ in SR1 and $35 \pm 21\,\%$ in SR4. The large statistical uncertainty in SR4 is due to low MC statistic. Because of this, the comparison is repeated using the 7 TeV ALPGEN MC samples, weighted to simulate 8 TeV collisions.[2] In all the SRs the results from SHERPA and from the ALPGEN 7 TeV re-weighted samples are compatible. The two comparisons

[2]The weights take into account the change in x of the in-going partons between 7 and 8 TeV collisions, and the consequent variation in PDF.

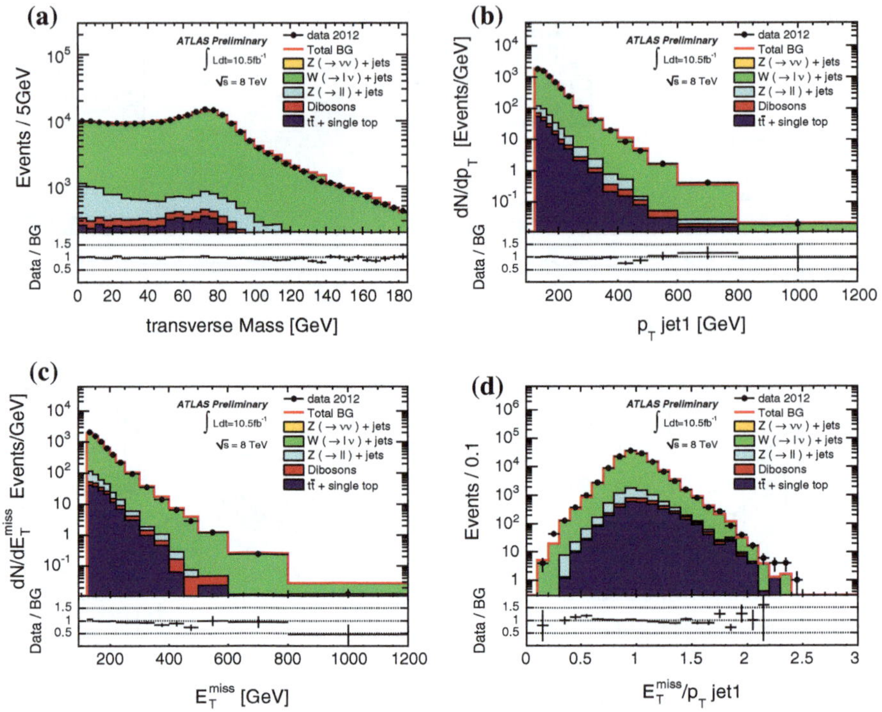

Fig. 4.3 Kinematic distributions of the events in the $W(\rightarrow \mu\nu)$ CR for the selection cuts of region 1 ($E_{\mathrm{T}}^{\mathrm{miss}}$, jet1 $p_{\mathrm{T}} > 120$ GeV)

are presented in Fig. 4.5. In the first three SRs the difference between the results from ALPGEN and SHERPA does not show a clear dependency with increased thresholds on p_{T} and $E_{\mathrm{T}}^{\mathrm{miss}}$. In SR4 the statistical uncertainties dominate so that is difficult to compare the results from the two MC generators. As for the 7 TeV analysis, an uncertainty of 3 % is considered for the four SRs.

The systematic uncertainties due to jet energy resolution and scale are considered. The latter is parametrized as a function of jet p_{T}, η, the number of reconstructed vertex, and μ (average number of interactions per bunch crossing). The jet uncertainties are then propagated to the $E_{\mathrm{T}}^{\mathrm{miss}}$. The total signal yield uncertainty due to jet and $E_{\mathrm{T}}^{\mathrm{miss}}$ is about 0.5 % in SR1 and grows up to 8 % in SR4. The uncertainty on the lepton identification, and momentum scale and resolution translates in a variation the BG estimates of about 0.8 % in SR1 to 1.4 % in SR4. The uncertainty related to the subtraction of BGs in the CRs is estimated varying the top and di-boson contributions by 20 % and translates in an uncertainty of the total Z/W+jet BG of about 1 % for all SRs. Finally, the uncertainty on the trigger efficiency results in an change of 0.2 % on the BG estimation of SR1. For the rest of the SRs, the trigger selection is fully efficient, therefore no uncertainty needs to be considered. Table 4.2 summarizes the systematic uncertainties on the total Z/W+jet BG estimation.

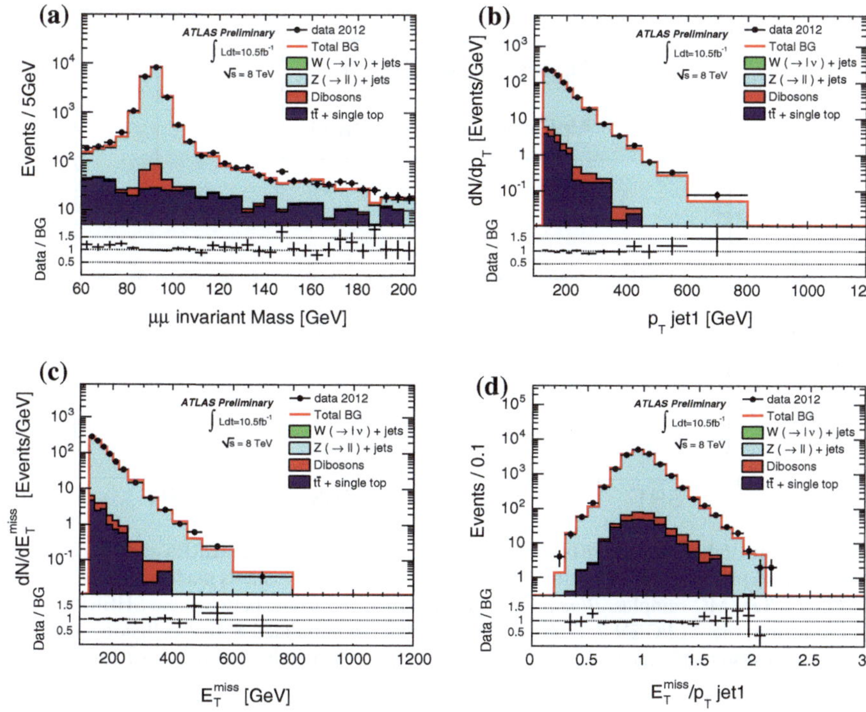

Fig. 4.4 Kinematic distributions of the events in the $Z(\rightarrow \mu\mu)$ CR for the selection cuts of region 1 ($E_{\mathrm{T}}^{\mathrm{miss}}$, jet1 p_{T} >120 GeV)

Table 4.1 Summary of the CR normalization factors with their corresponding statistical uncertainties from data and MC in the various CRs.

	Region 1	Region 2	Region 3	Region 4
$W \rightarrow \mu\nu$ CR	1.01 ± 0.01	0.94 ± 0.02	0.83 ± 0.05	1.10 ± 0.20
$Z \rightarrow \mu\mu$ CR	0.97 ± 0.01	0.93 ± 0.04	0.98 ± 0.12	1.40 ± 0.47
Electron inclusive CR	0.93 ± 0.01	0.89 ± 0.03	0.85 ± 0.08	0.91 ± 0.23

These factors are used for the normalization of the Z/W+jet MC samples in Figs. 4.2, 4.3 and 4.4

4.4.2 *Multi-jet Production*

The BG contamination from multi-jet events is estimated with the same technique as in the analysis carried out with 7 TeV data (see Sect. 3.5.2). For each SR we define two corresponding CRs in order to select di-jet and tri-jet events. The di-jet CR is defined by the same selection of the SR, but requiring a second jet with $\Delta\phi(jet2, E_{\mathrm{T}}^{\mathrm{miss}}) < 0.5$. The tri-jet CR is defined by the same requirements of the corresponding SR, but requiring a third jet with $\Delta\phi(jet3, E_{\mathrm{T}}^{\mathrm{miss}}) < 0.5$. Figure 4.6 shows both $\Delta\phi$ distributions before the cut on these variables.

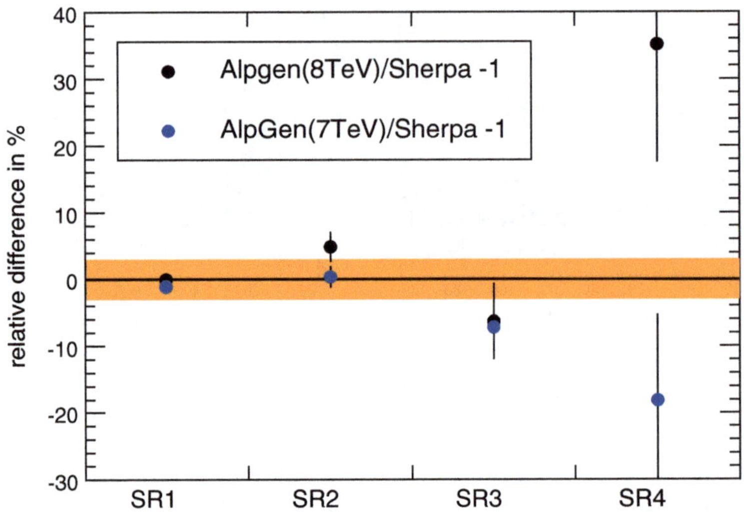

Fig. 4.5 Detail of the comparison between the total Z/W+jet BG using SHERPA and ALPGEN MC samples. *Black points* show the relative difference between the estimations made with SHERPA and ALPGEN 8 TeV MC samples. *Blue points* show the relative change between the SHERPA results and those from the ALPGEN 7 TeV samples, weighted to simulated 8 TeV collisions. Based on these result a systematic uncertainty of 3 % is chosen (represented by the *shadowed band*)

Table 4.2 Summary of the systematic uncertainties on the total Z/W+jet background

Systematic uncertainties				
	Region 1	Region 2	Region 3	Region 4
MC modeling (%)	3.0	3.0	3.0	3.0
JES and E_T^{miss} $(^{up}_{down})$ (%)	$^{+0.3}_{+0.5}$	$^{+0.1}_{-0.4}$	$^{+2.8}_{-0.5}$	$^{-7.8}_{-7.8}$
JER (%)	0.1	0.2	0.3	0.3
Lepton identification efficiency (%)	0.8	1.1	1.4	1.4
Background subtraction (%)	1.0	0.7	0.7	0.8
Trigger (%)	0.2	–	–	–
Luminosity (%)	–	–	–	–

Figures 4.7 and 4.8 show the distributions of the E_T^{miss} and p_T of the jets in the CRs corresponding to SR1. In the figures, the QCD MC samples are normalized to the number of events in data, which results in a factor 0.90 for the di-jet CR, and 0.82 for the tri-jet CR after the samples are normalized to the PYTHIA LO cross section.[3]

[3] For the 7 TeV analysis, the normalization factors were 0.76 and 0.78 respectively for the di-jet CR and tri-jet CR. The differences between the two analyses are expected because of the change in pile-up and detector conditions.

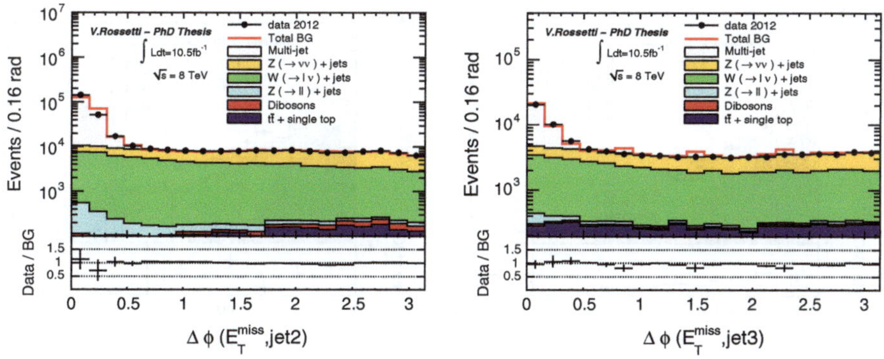

Fig. 4.6 The plot on the *left-hand side* shows the distribution of $\Delta\phi(\text{jet2}, E_{\text{T}}^{\text{miss}})$ in a di-jet configuration, while the plot on the *right-hand side* shows $\Delta\phi(\text{jet3}, E_{\text{T}}^{\text{miss}})$ in events with a tri-jet configuration. Both plots are done with the selection requirements corresponding to SR1 (leading jet p_{T} and $E_{\text{T}}^{\text{miss}} > 120$ GeV)

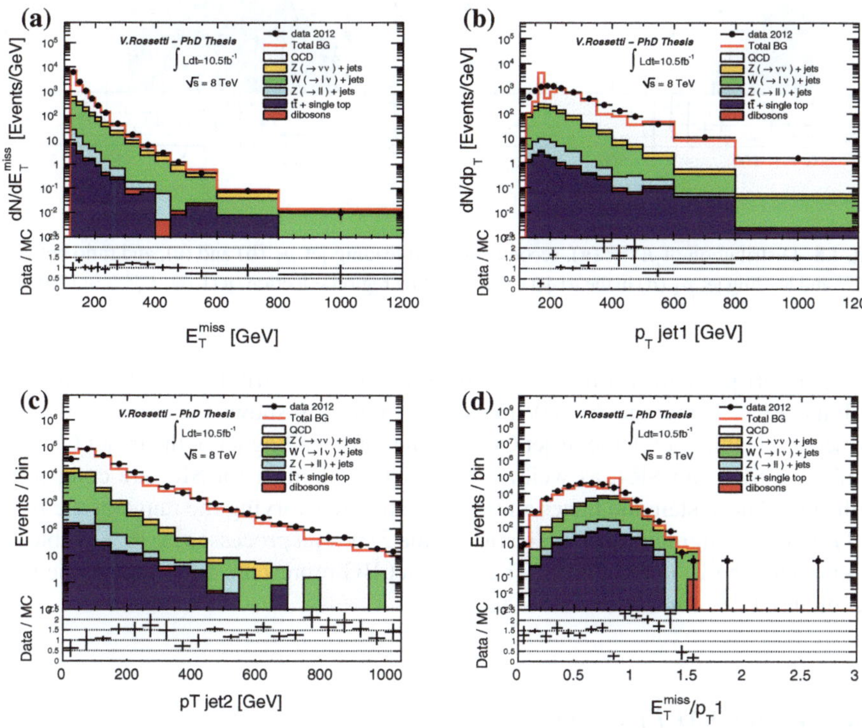

Fig. 4.7 Distribution of $E_{\text{T}}^{\text{miss}}$ (**a**), leading jet p_{T} (**b**), second leading jet p_{T} (**c**), ratio between $E_{\text{T}}^{\text{miss}}$ and leading jet p_{T} (**d**) in the di-jet CR for the first region selection (jet1 p_{T}, $E_{\text{T}}^{\text{miss}} > 120$ GeV)

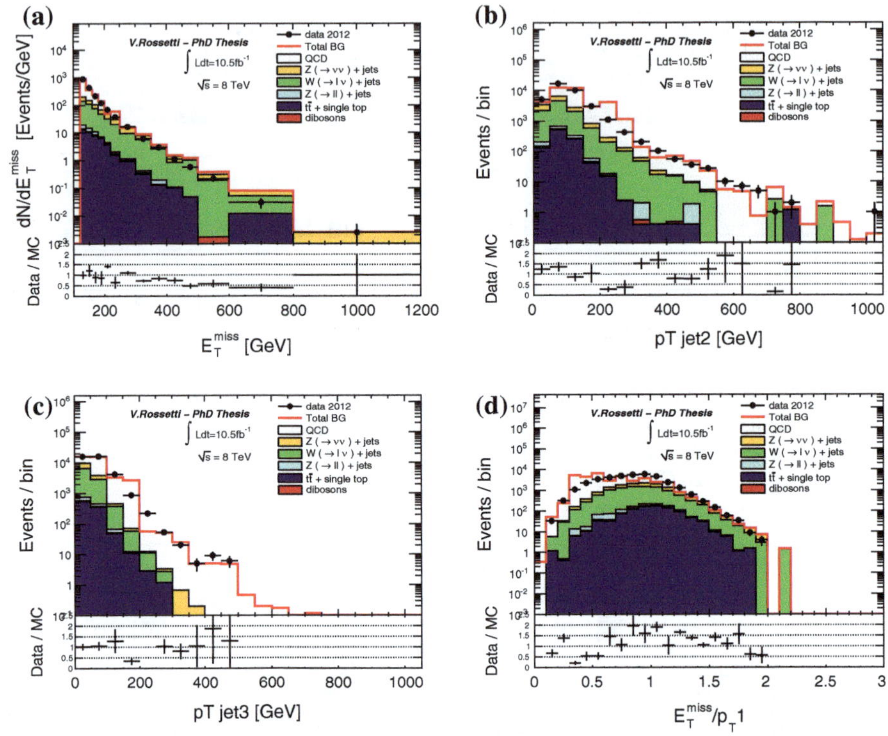

Fig. 4.8 Distribution of $E_{\mathrm{T}}^{\mathrm{miss}}$ (**a**), second and third leading jet p_{T} (**b**) and (**c**), and ratio between $E_{\mathrm{T}}^{\mathrm{miss}}$ and leading jet p_{T} (**d**) in the tri-jet CR for the first region selection (jet1 p_{T}, $E_{\mathrm{T}}^{\mathrm{miss}} > 120\,\mathrm{GeV}$)

The multi-jet BG is estimated extrapolating the p_{T} distribution of the jet aligned with the $E_{\mathrm{T}}^{\mathrm{miss}}$ below the 30 GeV threshold. Figure 4.9 shows the extrapolations of the second leading jet in the di-jet CR, and the third leading jet in the tri-jet CR. The results for SR3 and SR4, as well as the tri-jet component for SR2 are compatible with zero. The systematic uncertainties are estimated varying the range of the fit by 10 GeV and changing the normalization of the Z/W+jet processes by 10 %. Table 4.3 summarizes the BG predictions. The multi-jet BG represents a fraction of the total BG of about 2 % for SR1, and 1 % for SR2.

4.4.3 Other SM Processes

As for the 7 TeV analysis, the BG contributions from single top, $t\bar{t}$ and di-boson production are based on pure MC predictions. Results are shown in Table 4.4. Top and di-boson production represent a fraction of the total BG around 1 % for the four SRs. Systematic uncertainties on the MC predictions are estimated on $E_{\mathrm{T}}^{\mathrm{miss}}$/JES/JER,

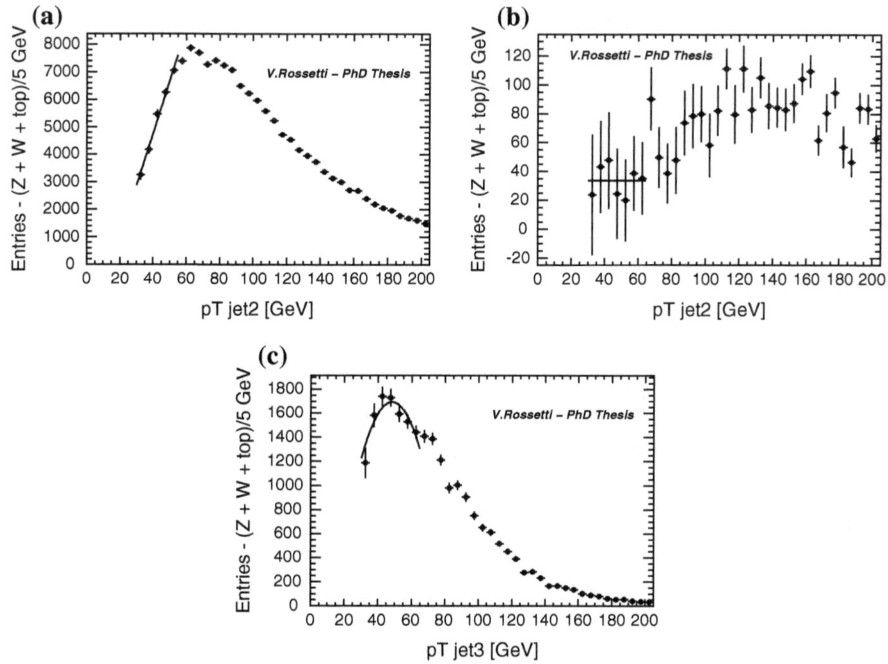

Fig. 4.9 Distribution of the second jet p_{T} in the di-jet events with the extrapolation to the SR1 using linear fit (**a**), and a fit to a constant for SR2 (**b**). The extrapolation of the third leading jet in the tri-jet CR corresponding to SR1 (**c**) is done with a second degree polinomial

Table 4.3 Results of the QCD background estimation

	SR1	SR2	SR3	SR4
Di-jet BG	4200±70±3500	200 ± 15 ± 200	–	–
Tri-jet BG	2200±50±2000	–	–	–
Total multi-jet BG	6400±90±5500	200 ± 15 ± 200	–	–

lepton identification, luminosity, trigger, cross section and PDF choice. The total uncertainty for both top and di-boson production is about 15 % for all SRs, and a conservative uncertainty of 20 % is adopted for all SRs.

4.4.4 Non-collision Background

In order to estimate the non-collision BG, we cannot use the same technique used for the 7 TeV dataset, that was based on data in un-paired bunch crossings. This is

Table 4.4 Summary of background estimations and total number of observed events in 10.5 fb^{-1} of data at $\sqrt{s} = 8$ TeV for the four mono-jet signal regions

	Background Predictions ± (stat.data)± (stat.MC) ± (syst.)			
	Signal region 1	Signal region 2	Signal region 3	Signal region 4
$Z(\rightarrow \nu\nu)$+jets	$173600 \pm 500 \pm 1300 \pm 5500$	$15600 \pm 200 \pm 300 \pm 500$	$1520 \pm 50 \pm 90 \pm 60$	$270 \pm 30 \pm 40 \pm 20$
$W(\rightarrow \tau\nu)$+jets	$87400 \pm 300 \pm 800 \pm 3700$	$5580 \pm 60 \pm 190 \pm 300$	$370 \pm 10 \pm 40 \pm 30$	$39 \pm 4 \pm 11 \pm 2$
$W(\rightarrow e\nu)$ +jets	$36700 \pm 200 \pm 500 \pm 1500$	$1880 \pm 30 \pm 100 \pm 100$	$112 \pm 5 \pm 18 \pm 9$	$16 \pm 2 \pm 6 \pm 2$
$W(\rightarrow \mu\nu)$ +jets	$34200 \pm 100 \pm 400 \pm 1600$	$2050 \pm 20 \pm 100 \pm 130$	$158 \pm 5 \pm 21 \pm 14$	$42 \pm 4 \pm 13 \pm 8$
$Z(\rightarrow \tau\tau)$+jets	$1263 \pm 7 \pm 44 \pm 92$	$54 \pm 1 \pm 9 \pm 5$	$1.3 \pm 0.1 \pm 1.3 \pm 0.2$	$1.4 \pm 0.2 \pm 1.5 \pm 0.2$
$Z(\rightarrow \mu\mu)$+jets	$783 \pm 2 \pm 35 \pm 53$	$26 \pm 0 \pm 6 \pm 1$	$2.7 \pm 0.1 \pm 1.9 \pm 0.3$	–
Multi-jets	$6400 \pm 90 \pm 5500$	$200 \pm 20 \pm 200$	–	–
$t\bar{t}$ + single t	$2660 \pm 60 \pm 530$	$120 \pm 10 \pm 20$	$7 \pm 3 \pm 1$	$1.2 \pm 1.2 \pm 0.2$
Di-bosons	$815 \pm 9 \pm 163$	$83 \pm 3 \pm 17$	$14 \pm 1 \pm 3$	$3 \pm 1 \pm 1$
Non-collision background	$640 \pm 40 \pm 60$	$22 \pm 7 \pm 2$	–	–
Total background	$344400 \pm 900 \pm 2200 \pm 12600$	$25600 \pm 240 \pm 500 \pm 900$	$2180 \pm 70 \pm 120 \pm 100$	$380 \pm 30 \pm 60 \pm 30$
Data	350932	25515	2353	268

In cases where both data and MCs play a role in the statistical errors, the two contributions are shown separated as second and third uncertainties. The last quoted uncertainty is the systematic uncertainty

because for most of the 2012 data-taking the LHC was running with a setting without un-paired bunch crossings at the ATLAS interaction point.

The estimation of the non-collision BG is obtained from data using the measured timing distribution of the leading jet. BG events from beam halo muons are caracterized by "early" jets, with the reconstructed time lower than zero. The shape of the timing distribution for non-collision BG is reconstructed from a control data sample with relaxed jet cleanup cuts. The number of non-collision BG in the SRs N^{SR}_{NCB} is computed as:

$$N^{SR}_{NCB} = N^{SR}_{-10<t<-5} \times \frac{N^{NCB}}{N^{NCB}_{-10<t<-5}}, \tag{4.1}$$

where $N^{SR}_{-10<t<-5}$ and $N^{NCB}_{-10<t<-5}$ denote the number of events in the SR and in the non-collision control sample, respectively, with a leading jet in the range -10 ns $<$ t < -5 ns, and N^{NCB} is the total number of events in the non-collision control sample. This results into 640±40(stat.)±60(syst.) and 22 ± 7 (stat.) ± 2 (syst.)

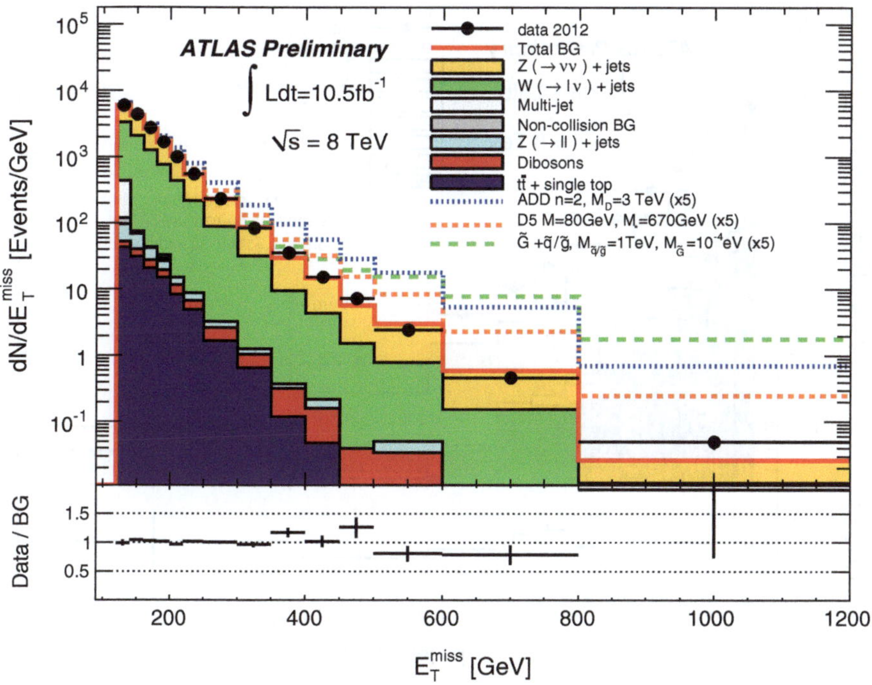

Fig. 4.10 $E_{\mathrm{T}}^{\mathrm{miss}}$ distribution of data and estimated BG in the signal region SR1. The *error bands* in the ratio plot reflect only the statistical uncertainty

non-collision BG events in the SR1 and SR2, respectively, while the non-collision background for SR3 and SR4 selections is negligible. The consistency between this method and the one described in the previous chapter, has been tested on the 7 TeV dataset, giving compatible results.

4.5 Results

In this section, the 8 TeV data (10.5 fb^{-1}) in the four SRs are compared with the SM predictions.[4] Figures 4.10 and 4.11 present the distributions of the $E_{\mathrm{T}}^{\mathrm{miss}}$ and leading jet p_{T} in the SR1. The error bars include only the statistical uncertainties from data and MC. For illustration, the impact of different ADD, WIMP, and GMSB signals are also included. A very good agreement between data and SM predictions is found. χ^2 tests performed on the two distributions of Figs. 4.10 and 4.11 (considering only statistical uncertainties) lead to χ^2 per degree of freedom values 1.9 and 1.1

[4]The results are documented in the ATLAS conference note ATLAS-CONF-2012-147 [1] and are considered preliminary.

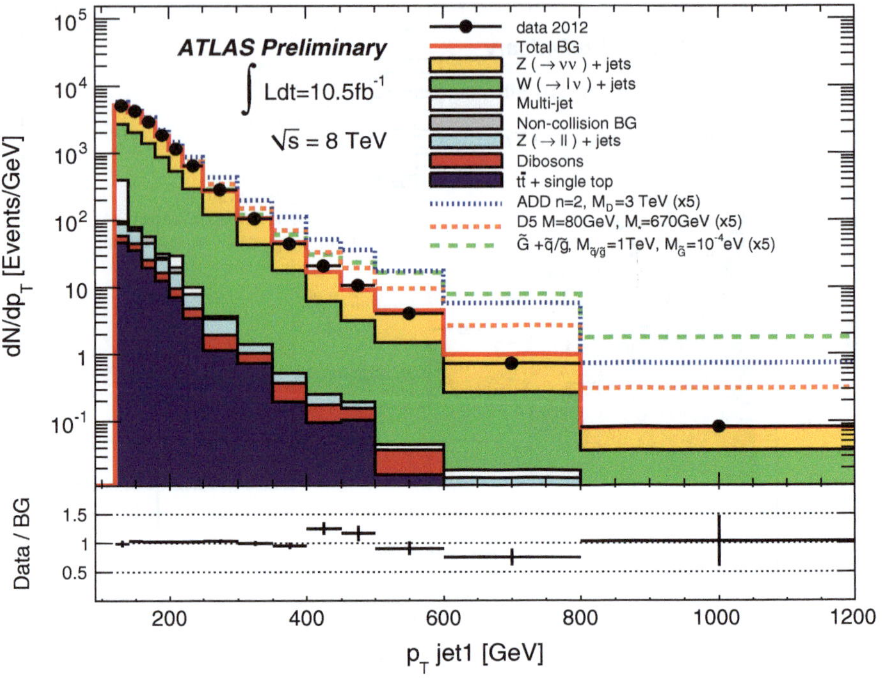

Fig. 4.11 Leading jet p_{T} distribution of data and estimated BG in the signal region SR1. The *error bands* in the ratio plot reflect only the statistical uncertainty

respectively. This shows a very good understanding of the SM BG processes (mainly Z/W+jets), in a large range of $E_{\mathrm{T}}^{\mathrm{miss}}$ and leading jet p_{T}.

Figure 4.12 shows other quantities related to the $E_{\mathrm{T}}^{\mathrm{miss}}$ and leading jet in SR1, testifying a good understanding of the $E_{\mathrm{T}}^{\mathrm{miss}}$ and the leading jet p_{T} spectrum, the jet angular distributions, and the ratio between the $E_{\mathrm{T}}^{\mathrm{miss}}$ and the jet p_{T}. Figure 4.13 shows distributions related to the second leading jet in SR1 again showing a good understanding of the jet radiation in the SM BG processes. Finally, Fig. 4.14 shows instead the $E_{\mathrm{T}}^{\mathrm{miss}}$ and leading jet p_{T} distributions for SR2, SR3 and SR4. A good agreement is found also in these distributions, despite the fact that the larger statistical uncertainties make difficult shape comparisons between data and SM predictions.

Table 4.4 reports the total observed number of events in data and lists the expected number of BG events and their statistical and systematic uncertainties for the four SRs. The uncertainty on the total SM predictions vary between 3.7 % in SR1 to 19 % in SR4. The data is compatible with SM background predictions, which, in SR3 and SR4, suffer from sizable statistical uncertainties, due mainly to lack of MC statistics. The comparison between the number of events in data with those of the estimated SM predictions is also shown graphically in Fig. 4.15.

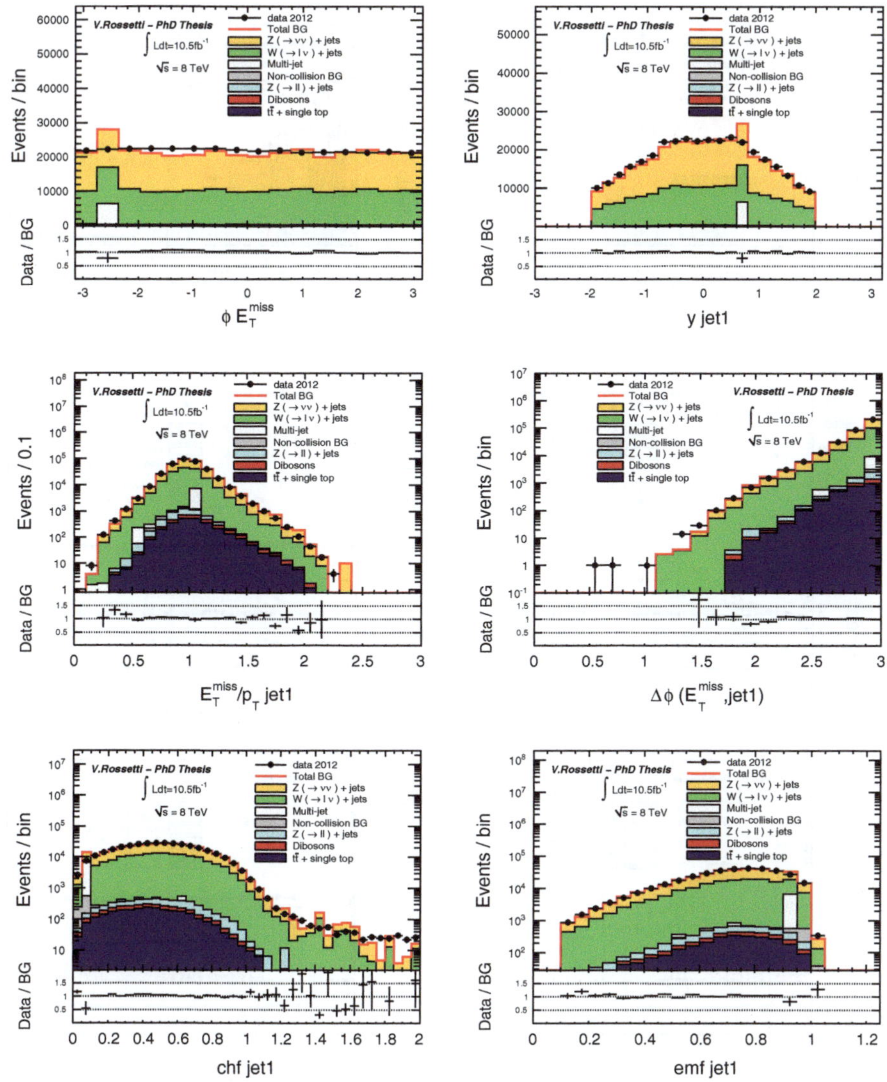

Fig. 4.12 Kinematic distributions of data and estimated BG in the signal region SR1. The *error bands* in the ratio plot reflect only the statistical uncertainty

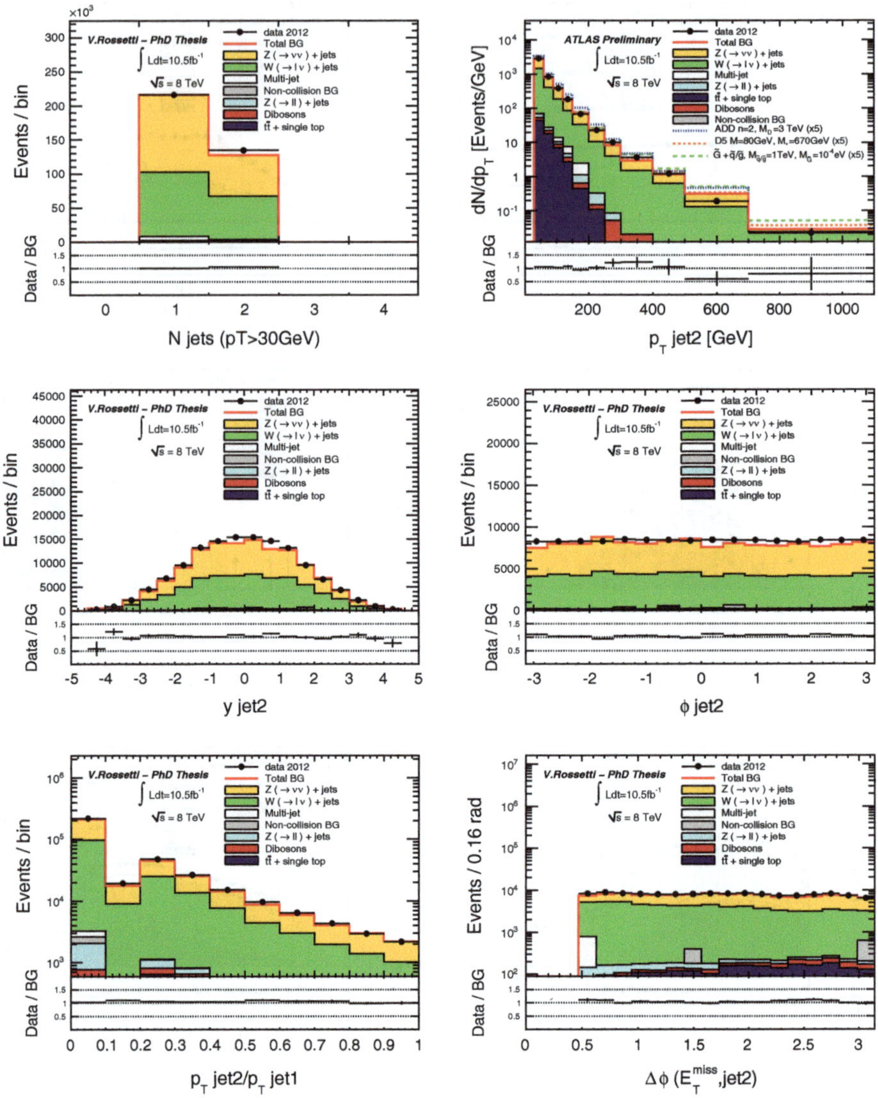

Fig. 4.13 Kinematic distributions of data and estimated BG in the signal region SR1. The *error bands* in the ratio plot reflect only the statistical uncertainty

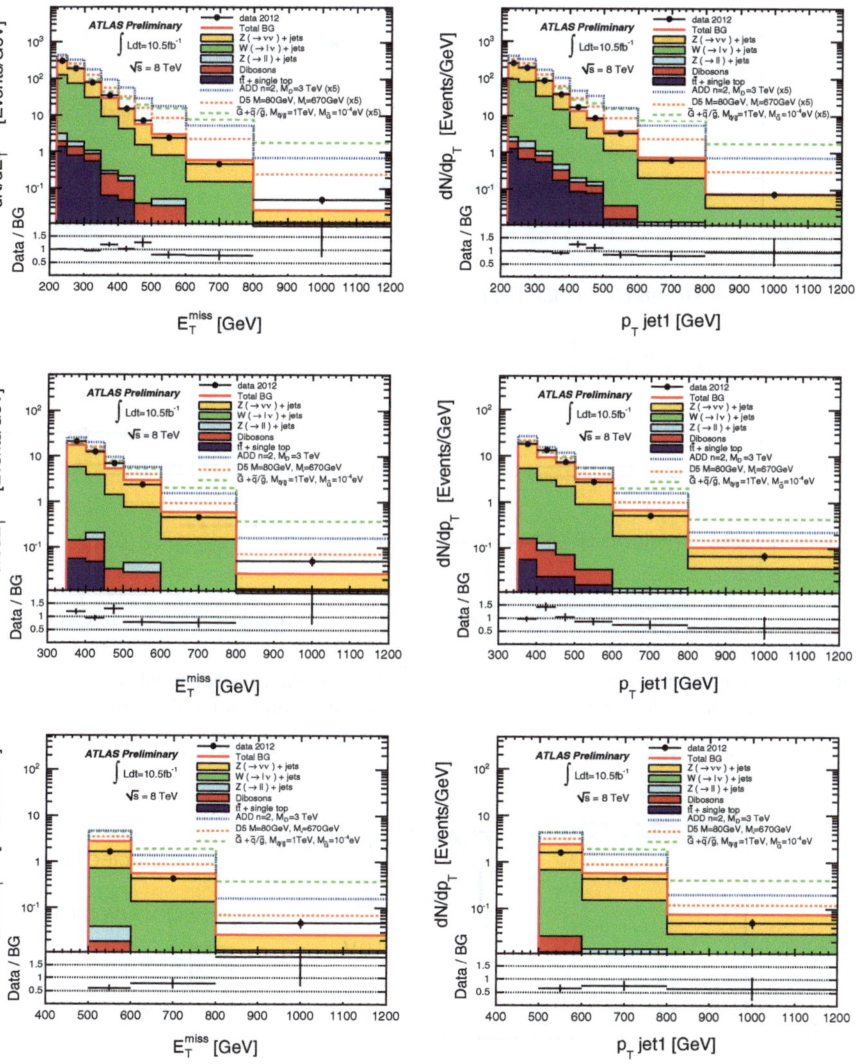

Fig. 4.14 Kinematic distributions of data and estimated BG in the signal region SR2, SR3 and SR4. The *error bands* in the ratio plot reflect only the statistical uncertainty

4.6 Interpretations

The 8 TeV results are translated into exclusion limits on physics beyond the SM, following the same procedure used to set limits in the 7 TeV analysis. The model-independent observed and expected limits on $\sigma_{vis} = \sigma \times A \times \epsilon$, are presented in Fig. 4.16 and Table 4.5. Values of σ_{vis} above 2.8 and 0.02 pb are excluded at 95 % CL for SR1 and SR4 selections, respectively. The results improve those from the 7 TeV

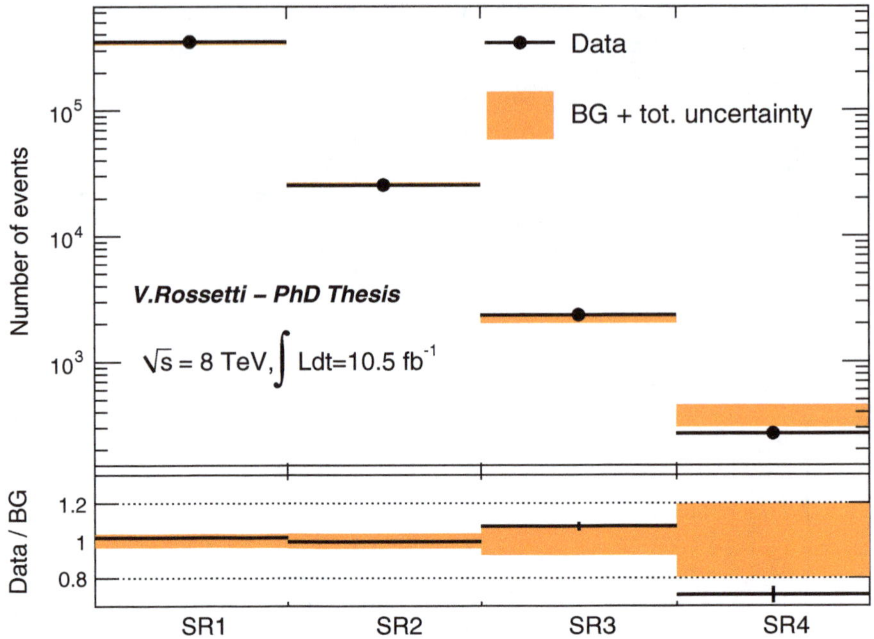

Fig. 4.15 Number of observed events in 10.5 fb^{-1} of data to the expected number of BG events in the four SRs. The *shadowed error band* shows the total uncertainty of the BG estimation

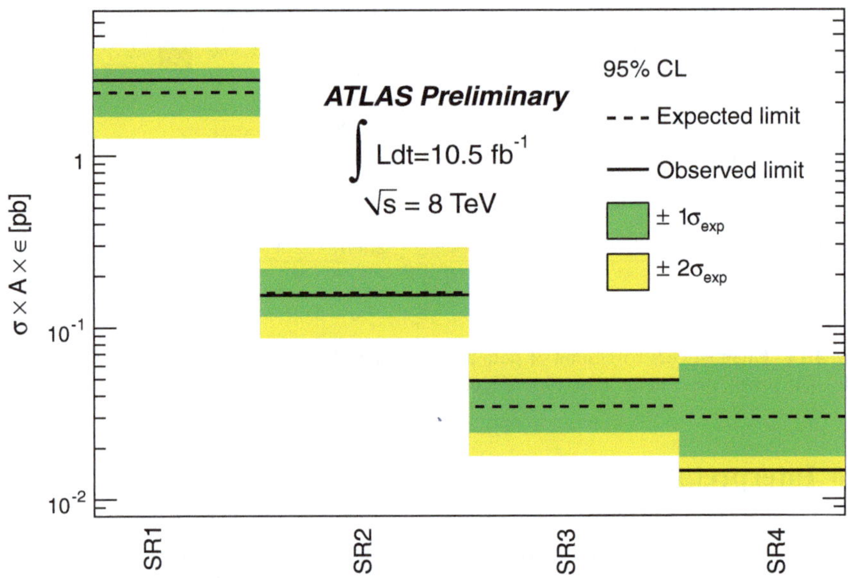

Fig. 4.16 The model-independent observed (*solid lines*) and expected (*dashed lines*) 95 % CL limits on $\sigma \times A \times \epsilon$ for the different SRs. The *shaded areas* around the expected limit indicate the expected $\pm 1\sigma$ and $\pm 2\sigma$ ranges of limits in the absence of a signal

Table 4.5 Model independent 90 and 95 % CL limits on the visible cross section on any process for physics beyond the SM for the four SRs

	Model independent limits on σ_{vis} (pb)			
	SR1	SR2	SR3	SR4
σ_{vis}^{obs} at 90 % CL	2.4	0.14	0.042	0.0011
σ_{vis}^{exp} at 90 % CL	2.0	0.14	0.029	0.0023
σ_{vis}^{obs} at 95 % CL	2.8	0.16	0.049	0.0015
σ_{vis}^{exp} at 95 % CL	2.4	0.16	0.035	0.0030

analysis although the improvement is not as large as naively expected, due to the large uncertainty on the BG prediction coming from the limited MC statistics.[5] In the following sections, we will study the interpretation of the results in terms of the GMSB scenario with light gravitino (as a new interpretation of the 8 TeV analysis), the ADD model, and the WIMP pair production. Final constraints on these models are based on SR3 results that gives the best expected limits.

4.6.1 Limits on Gravitino Production in GMSB Scenario

The results are interpreted in the context of a GMSB scenario with production of light gravitino associated with a squark or a gluino (see Sect. 1.4.3). We use a SUSY simplified model for which the gluino and squark decays lead to a gravitino and a gluon or a quark, respectively, producing a mono-jet final state. To set limits on this model, we consider gravitino masses $\mathrm{m}_{\tilde{G}}$ between 10^{-5} and 10^{-3} eV, and different configurations of squark and gluino masses, $\mathrm{m}_{\tilde{q}}$ and $\mathrm{m}_{\tilde{g}}$, up to 2.6 TeV. Figure 4.17 shows the phase-space mapped in $\mathrm{m}_{\tilde{q}} - \mathrm{m}_{\tilde{g}}$.

Figure 4.18 shows the $A \times \epsilon$ for the different squark/gluino mass combinations in SR3. It is worth to mention that the signal acceptance does not depend on $\mathrm{m}_{\tilde{G}}$ as it is very small. For $\mathrm{m}_{\tilde{g}} < \mathrm{m}_{\tilde{q}}$, the acceptance decreases, since the gluino+gravitino production dominates, resulting in a signal with slightly higher jet multiplicity.

As for the 7 TeV analysis, systematic uncertainties are divided into experimental and theoretical and they are treated differently in the limit setting. The dominant experimental uncertainty is due to jet/$E_{\mathrm{T}}^{\mathrm{miss}}$ scale and resolution and is estimated in the same way as for the Z/W+jets BG. It's effect on the signal yield ranges from 2 % and 16 % for different squark and gluino masses. The uncertainties on the luminosity and trigger efficiency result in an uncertainty on the signal yield of 3.6 % (for all SRs) and 1 % (only for SR1), respectively.

Theoretical uncertainties are estimated on the modeling of the initial and final state radiation (ISR and FSR), on the PDFs, and on the factorization and renormalization scales used. Uncertainties related to the ISR/FSR modeling are deter-

[5] At the time of writing this document the ATLAS collaboration has completed a massive MC production campaign that addresses this issue.

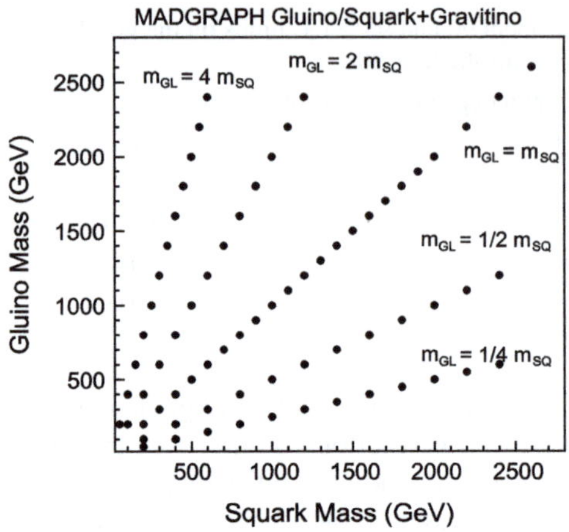

Fig. 4.17 Choices of the different signal points in the plane of the squark and gluinos masses, $m_{\tilde{q}}$ and $m_{\tilde{g}}$

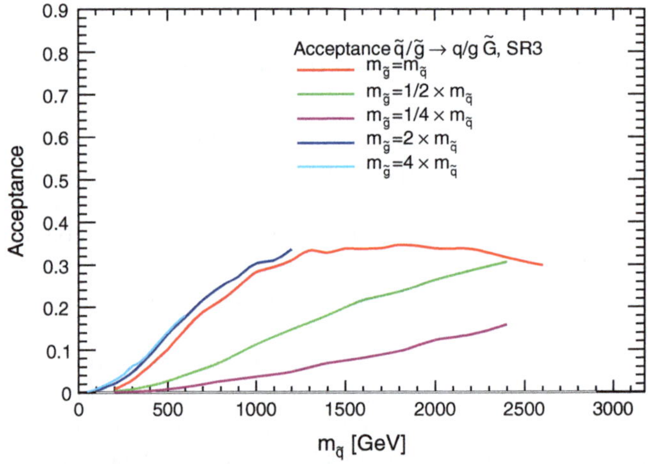

Fig. 4.18 $A \times \epsilon$ for the gravitino plus squark/gluino production in SR3 for different combinations of the squark and gluino masses

mined modifying the parameters that control the parton shower. This translates into a 5–10 % uncertainty on the signal yields, depending on the squark and gluino masses. Systematic uncertainties due to PDFs result in uncertainties on the signal yields that range from 5 %, for squark and gluino masses of 50 GeV, up to 60 % for masses of 2.6 TeV, reflecting the large PDF uncertainty at high x and high Q^2. Finally, variations of the renormalization and factorization scales by factors of two and one-half introduce a 15–35 % uncertainty on the signal yields, again typical for a LO calculation. The theoretical uncertainties as a function of the squark mass for different gluino masses are shown in Fig. 4.19.

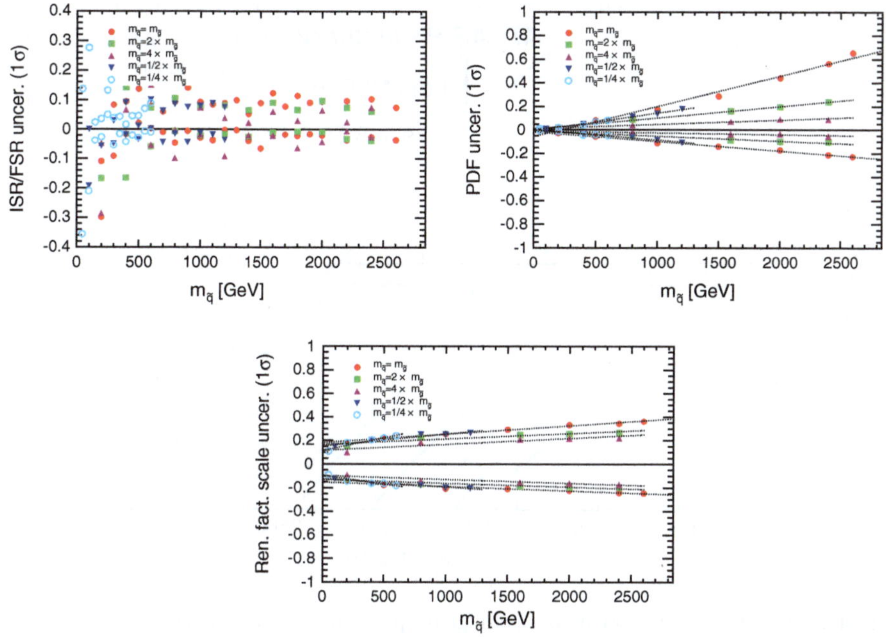

Fig. 4.19 Systematic uncertainties on the gravitino $\sigma \times A \times \epsilon$ for SR3 due to modeling of the ISR and FSR (*top left*), to PDFs (*top right*), and factorization and renormalization scales (*bottom*)

Figure 4.20 presents the signal $\sigma \times A \times \epsilon$ in the degenerate case ($m_{\tilde{g}} = m_{\tilde{q}}$) for different values of $m_{\tilde{G}}$. The expected and observed model-independent limits at 95 % CL are also shown, and the effect of varying the BG prediction by 1σ (2σ) of its total uncertainty is shown by the green (yellow) band around the expected limit.

Results are translated into exclusion limits on $m_{\tilde{G}}$, as a function of $m_{\tilde{g}}$ and $m_{\tilde{q}}$. Following the same procedure used in the 7 TeV analysis, the limits are calculated including all but the theoretical uncertainties on the signal. Figure 4.21 presents the 95 % expected and observed limits on the $m_{\tilde{G}}$–$m_{\tilde{q}}$ plane for degenerated squark and gluinos. gravitino masses below 10^{-4} eV are excluded at 95 % CL for $m_{\tilde{g}} = m_{\tilde{q}} = 500$ GeV. At $m_{\tilde{g}} = m_{\tilde{q}} = 1.7$ TeV gravitino masses below 4×10^{-5}eV are excluded. The effect of the theoretical uncertainties is illustrated by recomputing the observed limits for a signal yield varied by $-1\ \sigma_{THEO}$ (total theoretical uncertainty on the signal). This varies the constraints on $m_{\tilde{G}}$ by 10–20 %. Finally, the figures also indicate the region where the partial width for the gluino and squark to decay into a gravitino and a parton becomes more than 25 % of the mass. In this region the narrow width approximation is considered inappropriate since other decay channels should be considered.

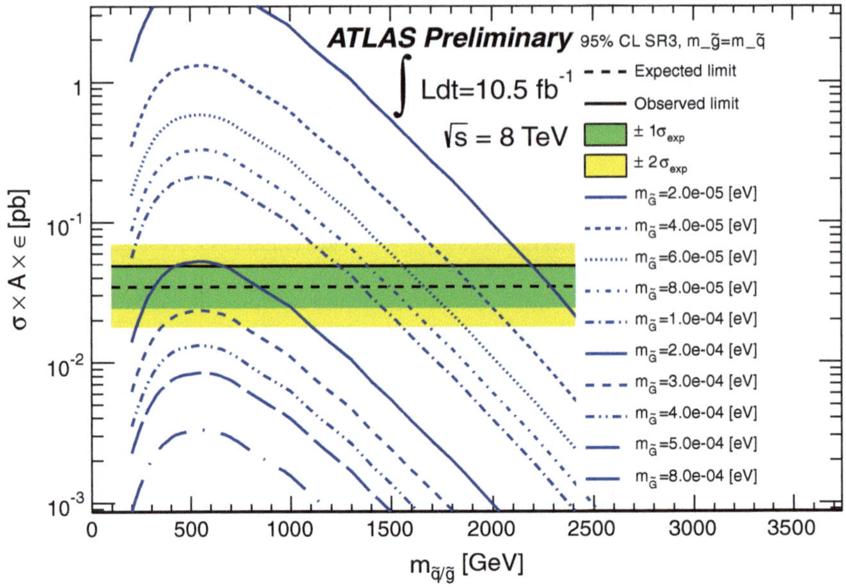

Fig. 4.20 $\sigma \times A \times \epsilon$ for the gravitino+squark/gluino production as a function of the squark/gluino mass in the case of degenerate squark and gluinos. Different values for the gravitino mass are considered and the predictions are compared with model-independent limits

Figure 4.22 shows the expected and observed limits calculated in the same way for gluino masses that are 1/4, 1/2, 2 and 4 times the squark mass. gravitino masses below 4×10^{-4} eV are excluded at 95 % CL in the case of $m_{\tilde{g}} = 2$ TeV and $m_{\tilde{q}} = 500$ GeV. For $m_{\tilde{g}} = 500$ GeV and $m_{\tilde{q}} = 2$ TeV values of $m_{\tilde{G}}$ lower then 10^4 eV are excluded.

As mentioned in Sect. 1.4.3, the gravitino mass is related to the scale of the SUSY breaking $\sqrt{\langle F\rangle}$ through the equation:

$$\langle F\rangle = \sqrt{3} m_{\tilde{G}} \bar{M}_{pl}. \tag{4.2}$$

Therefore, a lower bound of $m_{\tilde{G}} > 10^4$ eV translates into a constraint on $\sqrt{\langle F\rangle} > 645$ GeV. This exclusion extends significantly the previous one from LEP experiments ($\sqrt{\langle F\rangle} = 240$ GeV) [7].

Overall, the exclusion limits presented in this section largely extend the previous bounds from LEP [7] and Tevatron [8] (that excluded gravitino masses up to 1.37×10^{-5} eV), and constitute the best limits to date on the gravitino mass and the scale of the SUSY breaking in this scenario. The LEP/Tevetron results consider the pair production of gravitinos in a scenario with very large squark/gluino masses (above 2 TeV), whereas the limits presented in this section are based on the production of a gravitino with a squark/gluino, that then decays to a second gravitino and a parton, and fully use SUSY matrix elements.

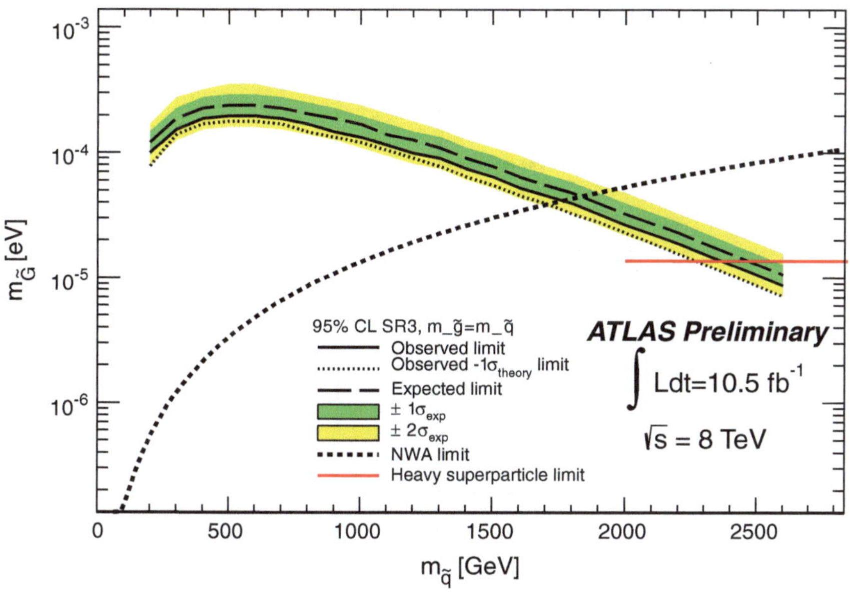

Fig. 4.21 Observed (*solid line*) and expected (*dashed line*) 95 % CL lower limits on the gravitino mass as a function of the squark mass for degenerate squark/gluino masses. The *dotted line* indicates the impact on the observed limit of the ± 1σ LO theoretical uncertainty. The *shaded bands* around the expected limit indicate the expected $\pm 1\sigma$ and $\pm 2\sigma$ ranges of limits in the absence of a signal. The *dashed-dotted line* defines the validity of the narrow-width approximation (see body of the text). The *solid red line* denotes the current limit from LEP on the gravitino mass assuming very heavy squarks/gluino

4.6.2 Limits on Graviton Production in ADD Model

Results are translated into 95 % CL limits on the parameters of the ADD model as in the 7 TeV analysis. The experimental and theoretical systematic uncertainties on the ADD signal are estimated for each SR and for number of extra dimensions n ranging from 2 to 6. The systematic uncertainties are listed in Table 4.7.

Figure 4.23 shows the predicted LO visible cross sections in SR3 for $n = 2$ and 6 as a function of M_D, and compared with the model-independent limits. Table 4.6 reports the 95 % CL exclusion limits on M_D based on LO predictions. Although the expected limits on M_D improved by ~15 % compared on the 7 TeV ones, the observed constraints do not supersede those obtained in the previous analysis (except in the case of $n = 6$).[6]

[6]As previously mentioned, these results are affected by large statistical fluctuations in the MC samples employed for the BG prediction (Table 4.7).

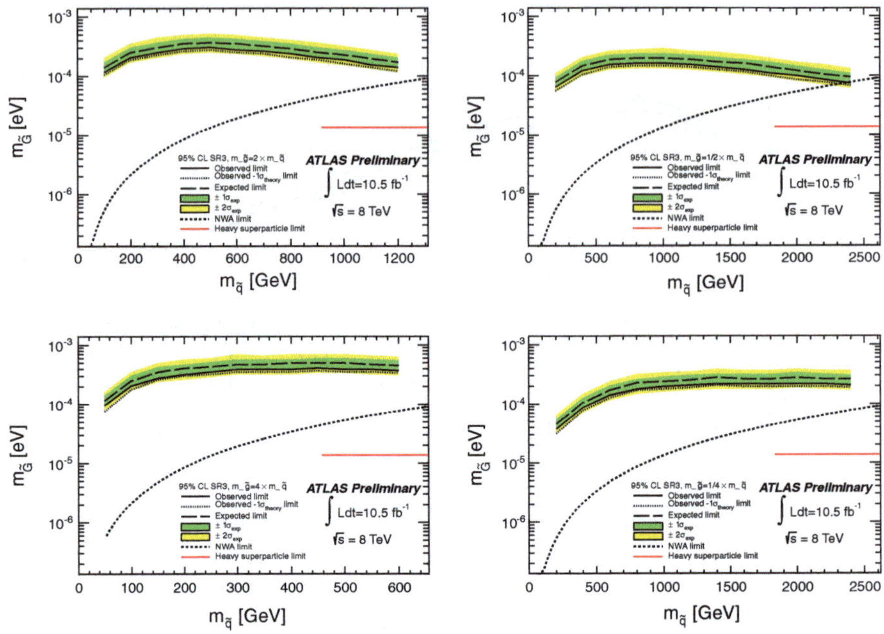

Fig. 4.22 Observed (*solid line*) and expected (*dashed line*) 95 % CL lower limits on the gravitino mass as a function of the squark mass for non-degenerate squark/gluino masses and different squark/gluino mass configurations. The *dotted line* indicates the impact on the observed limit of the $\pm 1\sigma$ LO theoretical uncertainty. The *shaded bands* around the expected limit indicate the expected $\pm 1\sigma$ and $\pm 2\sigma$ ranges of limits in the absence of a signal. The *dashed-dotted line* defines the validity of the narrow-width approximation (see body of the text). The *solid red line* denotes the current limit from LEP on the gravitino mass assuming very heavy squarks/gluino

Table 4.6 95 % CL expected and observed lower limits on M_D for different numbers of extra dimensions n, and in SR3

95 % CL limits on M_D		
n	exp (TeV)	obs (TeV)
2	4.24+0.39−0.36	3.88+0.32−0.42
3	3.39+0.46−0.24	3.16+0.21−0.29
4	3.00+0.20−0.16	2.84+0.16−0.27
5	2.78+0.15−0.13	2.65+0.16−0.27
6	2.69+0.11−0.11	2.58+0.13−0.23

The $\pm 1\sigma$ due to theoretical systematic uncertainties on expected and observed limits are also quoted

Table 4.7 Relative systematic uncertainties from each source, along with the total relative systematic and statistical uncertainties, (in %), on the ADD signal yield

n	PDF	ISR/FSR	Fact.Ren.	JES/$E_{\mathrm{T}}^{\mathrm{miss}}$	Lumi.	Trig.	MC Sta.	Tot.Syst
Signal region 1								
2	5.1	6.0	37	6.1	3.6	1.0	0.87	38
3	8.3	7.6	38	4.8	3.6	1.0	0.84	40
4	14	6.9	38	4.8	3.6	1.0	0.83	41
5	22	5.4	33	3.8	3.6	1.0	0.83	41
6	31	4.8	31	4.5	3.6	1.0	0.81	45
Signal region 2								
2	7.5	2.7	43	7.2	3.6	0	1.6	44
3	11	7.9	41	5.9	3.6	0	1.4	43
4	17	3.3	38	5.3	3.6	0	1.3	43
5	26	5.3	40	5.6	3.6	0	1.2	49
6	35	3.2	31	4.9	3.6	0	1.2	47
Signal region 3								
2	10	6.7	35	10	3.6	0	2.8	39
3	14	7.1	34	8.4	3.6	0	2.3	39
4	21	2.7	33	7.3	3.6	0	1.9	40
5	31	6.2	43	6.9	3.6	0	1.8	54
6	39	2.8	35	7.0	3.6	0	1.7	53
Signal region 4								
2	13	8.7	33	14	3.6	0	5.3	40
3	19	12	28	11	3.6	0	3.8	38
4	24	6.0	38	9.8	3.6	0	3.0	47
5	35	14	39	6.0	3.6	0	2.7	55
6	45	9.6	42	8.1	3.6	0	2.6	63

Table 4.8 The 90 % (95 %) CL observed lower limits on the suppression scale M* as a function of WIMP mass m_χ

m_χ (GeV)	D5	D8	D11
80	731(704)	713(687)	309(301)
400	632(608)	535(515)	257(250)
1000	349(336)	250(240)	155(151)

All values correspond to the nominal observed limit excluding theoretical uncertainties. The signal region with the best expected limits, SR3, is quoted in all cases

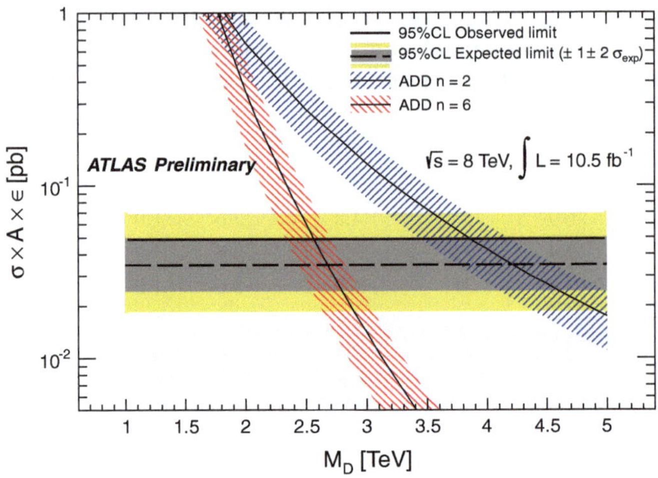

Fig. 4.23 The predicted ADD $\sigma \times A \times \epsilon$ for the SR3 selection as a function of MD for n = 2 and n = 6, where bands represent the uncertainty on the theory. For comparison, the model-independent observed (*solid line*) and expected (*dashed line*) 95 % CL limits on $\sigma \times A \times \epsilon$ are shown. The *shaded areas* around the expected limit indicate the expected $\pm 1\sigma$ and $\pm 2\sigma$ ranges of limits in the absence of a signal

4.6.3 Limits on WIMP Pair Production

The results are also translated into exclusion limits on WIMP pair production, considering the effective operators D5, D8 and D11, and WIMP masses $m_\chi = 80$, 400, and 1000 GeV. Experimental and theoretical systematic uncertainties on the WIMP signal are estimated as for the 7 TeV analysis, and are listed in Table 4.9, for different values of the WIMP mass m_χ. Since the simulated signal events with the operator D8 exhibit the same kinematic behavior as D5 and only differs in cross section, all systematic uncertainties of D5 are hence used for D8 too, and the difference in cross section is taken into account in the limit setting.

For $m_\chi = 80$ GeV, values of M* lower than 730 GeV are excluded at 90 % CL for the operator D5. Instead, for the operator D11 the lower bounds on M* are at 310 GeV. Figure 4.24 shows the 90 % CL limits on the scale suppression M* as a function of m_χ and for the three operators. Table 4.8 reports the 90 and 95 % CL limits on M* for the different operators and m_χ. Note that compared to the 7 TeV analysis, the limits of D5 and D8 are now about 10 % larger and hence more restrictive. For D11, the result does not improve the previous analysis (Table 4.9).[7]

[7] As for the limits on the ADD model, the improvement is not as large as one would naively expect from the increase of energy and integrated luminosity, because of the lack of sufficient MC statistics.

Table 4.9 Overview of WIMP signal relative uncertainties (percentage) for SR3 and SR4

Uncertainty	Operator	m_χ (GeV)	SR3	SR4
ISR/FSR	All	All	6.0	5.6
Jet matching scale	All	All	1.5	2.4
Fact.Renorm.Scale		80	9.5	12.6
	D5	400	6.4	5.9
		1000	13.4	13.4
		80	28.7	31.6
	D11	400	29.9	30.0
		1000	29.8	30.3
PDF		80	7.3	9.2
	D5	400	10.4	11.1
		1000	29.3	30.7
		80	24.6	25.1
	D11	400	39.3	39.3
		1000	87.9	87.9
JES/JER/E_T^{miss}(up/down)		80	11.0 / −6.6	16.0 / −8.9
	D5	400	8.8 / −4.7	10.0 / −4.3
		1000	8.5 / −6.6	10.0 / −9.1
		80	6.8 / −3.0	7.1 / −2.3
	D11	400	6.7 / −0.85	7.1 / −2.3
		1000	5.9 / 1.5	7.6 / −4.2
Luminosity	All	All	3.6	3.6
Trigger	All	All	0	0

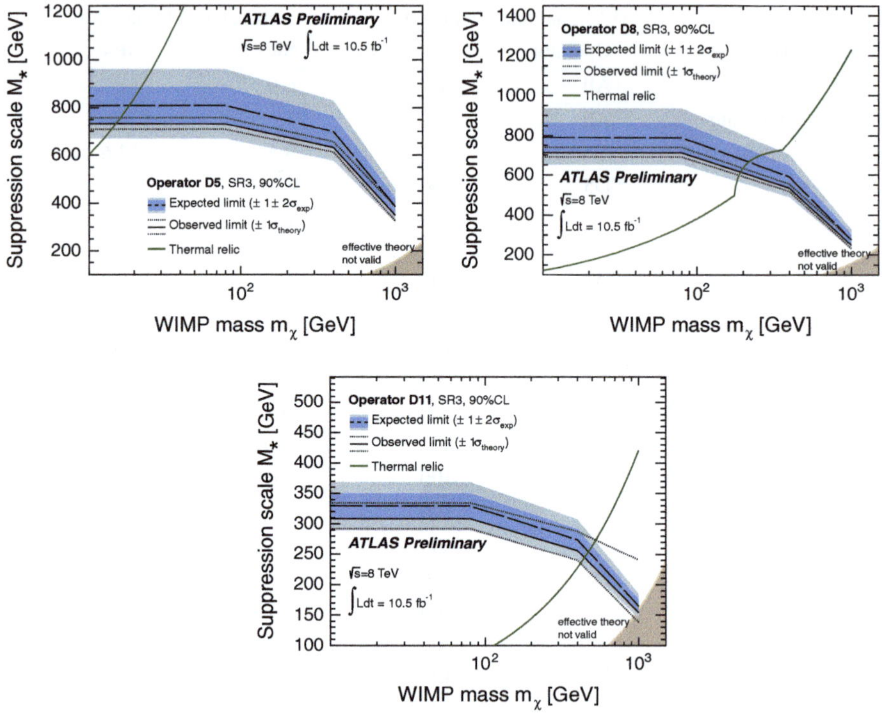

Fig. 4.24 The 90 % CL lower limits on M* for different masses of χ. Observed and expected limits including all but the theoretical signal uncertainties are shown as *dashed black* and *red solid lines*, respectively. The *grey* and *blue bands* around the expected limit are the ± 1 and 2σ variation expected from statistical fluctuations and experimental systematic uncertainties on SM and signal processes. The impact of the theoretical uncertainties is shown by the *thin red dotted* $\pm 1\sigma$ limit lines around the observed limit. The M* values at which WIMPs of a given mass would result in the required relic abundance are shown as rising *green lines* (taken from [9]), assuming annihilation in the early universe proceeded exclusively via the given operator. The *shaded light-grey* regions in the bottom right corners indicate where the effective field theory approach breaks down. The plots are based on the best expected limits, which correspond to SR3

References

1. ATLAS Collaboration, Search for new phenomena in monojet plus missing transverse momentum final states using 10/fb of pp collisions at $\sqrt{s} = 8$TeV with the ATLAS detector at the LHC (ATLAS-CONF-2012-147), July 2012
2. G. Aad et al., Search for dark matter candidates and large extra dimensions in events with a jet and missing transverse momentum with the ATLAS detector. JHEP **04** (2013)
3. Georges Aad et al., Search for new phenomena with the monojet and missing transverse momentum signature using the ATLAS detector in $\sqrt{s} = 7$TeV proton-proton collisions. Phys. Lett. B **705**, 294–312 (2011). doi:10.1016/j.physletb.2011.10.006
4. Search for new phenomena in monojet plus missing transverse momentum final states using 1 fb^{-1} of pp collisions at $\sqrt{s} = 7$TeV with the atlas detector (ATLAS-CONF-2011-096), Jul 2011
5. K. Mawatari, Y. Takaesu, HELAS and madGraph with goldstinos. Eur. Phys. J. C **71**, 1640 (2011)
6. ATLAS Collaboration, Search for new phenomena with mono-jet plus missing transverse energy signature in pp collisions at $\sqrt{s} = 8$TeV with the atlas detector (ATL-COM-PHYS-2012-1211), October 2012
7. 183-208 GeV LEP2 SUSY Working Group Collaboration, Single Photons. http://lepsusy.web.cern.ch/lepsusy/www/photons/single/singlepublicsummer04.html
8. M. Klasen, G. Pignol, New results for light gravitinos at hadron colliders: tevatron limits and lhc perspectives. Phys. Rev. D **75**, 115003 (2007)
9. Jessica Goodman, Masahiro Ibe, Arvind Rajaraman, William Shepherd, Tim M.P. Tait et al., Constraints on dark matter from colliders. Phys. Rev. D **82**, 116010 (2010). doi:10.1103/PhysRevD.82.116010

Chapter 5
Conclusions

This thesis presents results on the search for new phenomena in mono-jet events using proton–proton data collected by the ATLAS experiment at LHC collider with center-of-mass-energies of 7 TeV (in 2010 and 2011) and 8 TeV (in 2012). The results have been interpreted in the context of the production of Weak Interactive Massive Particles (WIMPs), a model with Large Extra Dimensions (ADD), and the production of light gravitinos in association with a squark or a gluino in Supersymmetry. The full 7 TeV dataset and nearly half of 8 TeV one have been used, and the results led to two papers [1, 2] and two public notes [3, 4] within the ATLAS collaboration.

The precise determination of the Z/W+jets background, that constitutes the dominant background contribution, is a central part of this thesis. A dedicated data-driven method is employed making use of data events in control regions. This translates in a significant reduction of the systematic uncertainties to the level of few percents. Other processes (QCD multi-jets and non-collision background) are estimated with dedicated data-driven methods. Finally, sub-leading contributions (top and di-boson production) are determined from Monte Carlo simulation.

In all cases, the data are in agreement with the Standard Model (SM) predictions, and the results are interpreted in the context of various scenarios for physics beyond the SM. For the ADD model, the fundamental Planck scale in $4+n$ dimensions M_D is constrained, and values of M_D lower than 4.37 TeV for $n = 2$, and lower than 2.69 TeV for $n = 6$, are excluded at 95 % confidence level (CL), challenging altogether the validity of this model.

Results are also interpreted in the context of WIMP pair production. An effective lagrangian is used to describe several types of interactions between WIMPs and SM particles (quarks or gluons). Different effective operators result in either spin-dependent or spin-independent interactions. Exclusion limits on the WIMP-nucleon cross section are then derived and compared with direct dark matter detection experiments. The ATLAS results give a unique access to WIMP masses $m_\chi < 10$ GeV (where direct detection suffers from kinematic suppression), and are particularly relevant for spin-dependent operators and in the case of WIMP-gluon interaction. Results are also translated into constraints on the WIMP–WIMP annihilation rate, and compared with results from indirect detection experiments.

V. Rossetti, *Search for Exotic Mono-jet Events*, Springer Theses,
DOI 10.1007/978-3-319-22225-7_5

Finally, the results are interpreted in the context of Gauge Mediated Supersymmetry Breaking with light gravitino. In this scenario, the associated production of gravitinos with a squark or gluino has been studied for different configurations of the squark and gluino masses, $m_{\tilde{q}}$ and $m_{\tilde{g}}$. In the case $m_{\tilde{q}} = m_{\tilde{g}} = 1.7$ TeV, gravitino masses below 4×10^{-5}eV are excluded at 95 % CL. This is used to infer a lower bound on the scale of the Supersymmetry breaking of $\sqrt{\langle F \rangle} = 645$ GeV, extending significantly the previous contraint from LEP.

During the year 2015, the LHC is resuming operations to provide p-p collisions up to $\sqrt{s} = 13$ –14 TeV, accessing a range of energies never studied before, and giving the opportunity for new exciting discoveries in the mono-jet final state.

References

1. G. Aad et al., Search for new phenomena with the monojet and missing transverse momentum signature using the ATLAS detector in $\sqrt{s} = 7$ TeV proton-proton collisions. Phys. Lett. **B705**, 294–312 (2011). doi:10.1016/j.physletb.2011.10.006
2. G. Aad et al., Search for dark matter candidates and large extra dimensions in events with a jet and missing transverse momentum with the ATLAS detector. JHEP **04** (2013)
3. Search for new phenomena in monojet plus missing transverse momentum final states using 1 fb^{-1} of pp collisions at $\sqrt{s} = 7$ TeV with the atlas detector (ATLAS-CONF-2011-096), July 2011
4. ATLAS Collaboration, Search for new phenomena in monojet plus missing transverse momentum final states using 10/fb of pp collisions at $\sqrt{s} = 8$ TeV with the ATLAS detector at the LHC (ATLAS-CONF-2012-147), July 2012

Appendix A
TileCal Commissioning with Cosmic Muons, Single Beam, and Collision Data

This appendix describes the commissioning and calibration of the ATLAS Tile hadronic calorimeter (TileCal). Until December 2009 the ATLAS detector went through a long period of commissioning with cosmic rays. This data was used to study the energy response and the inter-calibration among cells. Subsequently the LHC operated in special settings with single beams and delivered the so called "beam splash" events, that were used in TileCal for time calibration. Starting from December 2009, the LHC provided proton–proton collisions at different center of mass energies (900 GeV, 2.36 and 7 TeV). The early data was used for the commissioning of the ATLAS detector for the 2010 collision run. The studies described in this appendix are published in the Eur. Phys. J., in "Readiness of the ATLAS Tile Calorimeter" [1].

TileCal is divided into three cylindrical sections, whose axes coincide with the colliding beam axis (z axis). The cylindrical sections are referred to as long barrel (LB), that covers a region $|\eta| \lesssim 0.8$, and extended barrels (EB), covering approximately $0.8 < |\eta| < 1.6$. Each of the three sections is composed of 64 azimuthal segments, referred to as modules, subtending $\Delta\phi = 2\pi/64 \simeq 0.1$. The electronic readout is organized in four independent blocks, called "partitions": one for each EB, and two for the LB. Radially, the TileCal cells are organized in three layers called A (the most internal), BC (the middle one), and D (the external).

Calorimeter Response to Cosmic Muons

Since the interaction of muons with matter is well understood and well modeled, cosmic rays measurements were used to provide information about the response of the detector. The cosmic ray events used in this analysis were collected between September and October 2008. The trigger selection of the events relies on the RPC and TGC systems of the Muon Spectrometer (see Sect. 2.2.3). The analysis is based on about 1M triggered events. The tracks are reconstructed using the Inner Detector (see Sect. 2.2.1) information and are extrapolated through the volume of the ATLAS detector. For each muon track crossing a calorimeter cell, a path length $\mathrm{d}x$ is evaluated

V. Rossetti, *Search for Exotic Mono-jet Events*, Springer Theses,
DOI 10.1007/978-3-319-22225-7

as the distance between the entrance and the exit points. The effect of the magnetic field on the path reconstruction is found to be negligible, and the muon trajectory inside the cell is approximated with a straight line.

The events are required to have a reconstructed track in the Inner Detector with at least 8 hits in the Pixel and SCT systems. The transverse and longitudinal impact parameters are required to be $|d_0| < 380$ mm and $|z_0| < 800$ mm respectively. The events are further required to have the track momentum in the window 10 GeV $< p <$ 30 GeV. The lower limit was applied to minimize effects of multiple scattering. The upper cut restricts the muon radiative energy losses which otherwise cause considerable fluctuations in the deposited energy. A small fraction of events has the cell response compatible with the pedestal level although the cells should be hit by the muon. This happens because the particle actually crosses a neighboring cell and it is consistent with the expected deviation from a linear muon trajectory due to multiple scattering. In order to limit this effect, only tracks that are well within a module were selected by applying the cuts $|\Delta\phi_{in}| = |\phi_{cell} - \phi_{in}| < 0.045$ and $|\Delta\phi_{out}| = |\phi_{cell} - \phi_{out}| < 0.045$, where $\phi_{in}(\phi_{out})$ is the azimuthal angle of the entrance (exit) point of the muon trajectory in the cell. In order to remove residual noise contribution, the measured cell energy $\mathrm{d}E$ is required to be larger than $\mathrm{d}E_{cut} = 60\,\mathrm{MeV}$. Deposits of energy in a short path have a larger variation of the sampling fraction. In order to limit this effect only tracks with path length $\mathrm{d}x > 200$ mm were considered. The energy deposited by a muon track with a trajectory close to the vertical direction is badly measured in TileCal due to the strong sampling fraction variation caused by the vertical orientation of the scintillating tiles. To ensure more stable results, tracks are required to have a minimum cell path z component. The cut $z = |z_{in} - z_{out}| > 6\,\mathrm{cm}$ was applied, where z_{in} and z_{out} are the crossing points at the z cell surfaces.

At energies in the range between 10 and 30 GeV the dominant energy loss process is ionization and the energy lost is essentially proportional to the track path length (Fig. A.1-left). The response of the detector was then studied with the truncated mean of $\mathrm{d}E/\mathrm{d}x$, defined as the mean $\mathrm{d}E/\mathrm{d}x$ in which 1 % of the events in the high-energy tails of the distribution are removed. This quantity is less affected by the higher energy loss proceses (bremsstralung, δ-rays...) that can cause fluctuations on the mean $\mathrm{d}E/\mathrm{d}x$.

The uniformity of the cell response in the layer D (the most external) is shown in Fig. A.1-right. The cells considered are required to have at least 100 selected muon deposits. For the D layer 316 cells fulfill this requirement and represent 38 % of the total number of cells in this layer. The distribution is compared with MC simulated cosmic events, in which the calorimeter response is, by definition, equal in every cell. Since the MC shows an RMS compatible with that of data, it indicates that cells are well inter-calibrated within the layer. Figure A.2 shows the cell response, expressed in terms of truncated mean of $\mathrm{d}E/\mathrm{d}x$, as a function of ϕ and η for the layer A cells. The $\mathrm{d}E/\mathrm{d}x$ values are normalized on the average $\mathrm{d}E/\mathrm{d}x$ of all the cells of the layer. Data and MC show similar pattern and the dispersion of the data points is below 3 % for the explored range.

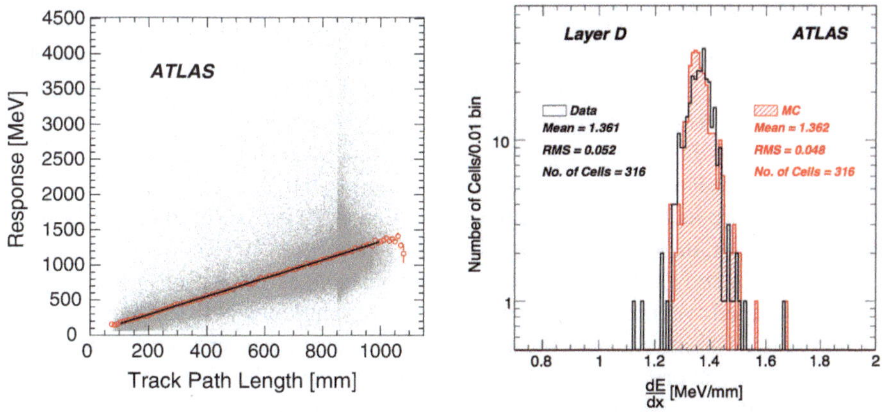

Fig. A.1 On the *left*, the response of the barrel module BC cells as a function of track path length is shown. A linear fit to the corresponding distribution of mean values is superimposed. The excess of events around the track path length of 840 mm (radial size of the BC cells of the barrel module) is a purely statistical effect, since most of the cosmic ray muons enter the calorimeter at small zenith angle. On the *right*, the distribution of the truncated mean $\mathrm{d}E/\mathrm{d}x$ per cell, shown for the radial layer D for data and MC

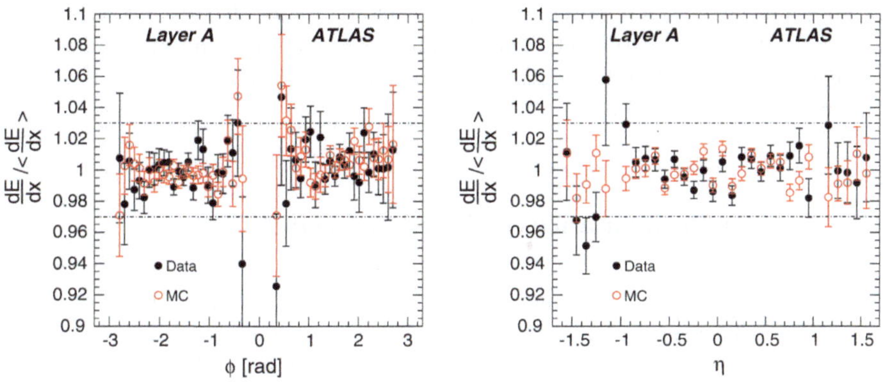

Fig. A.2 Uniformity of the A layer cell response to cosmic muons, expressed in terms of normalized truncated mean of $\mathrm{d}E/\mathrm{d}x$, as a function of azimuthal angle ϕ (*left*) and the pseudo-rapidity η (*right*)

Timing Calibration with "Splash" Events

To allow for optimal reconstruction of the energy deposited in the calorimeter the timing of the readout channels must be adjusted with a precision of $\sim$1 ns. These time offsets can be measured using the laser calibration system [1]. The timing precision for channels in the same module achieved with laser pulse on the photomultipliers (PMTs), can be as low as 0.6 ns for 99 % of the Tile Calorimeter readout channels. Nevertheless, the timing calibration with the laser system has some limitations as, for example, the knowledge of the propagation time in the optical fibers between the laser source to the PMTs, and the propagation time between the cells and the PMTs.

The inter-partition timing and global timing with respect to the rest of ATLAS were coarsely set using cosmic-ray data, and more accurately using the so called "splash" events. In these events the LHC proton beam hits a completely closed collimator placed on the beam line 140 m away from ATLAS interaction point. This settings produce O(10^5) particles (mainly muons) arrive simultaneously in the ATLAS detector, depositing in TileCal a total energy of $\sim 10^3$ TeV.

Figure A.3a shows the time of the energy deposits as measured in beam splash events, averaged over the full range of the azimuthal angle ϕ for all cells with the same z-coordinate. The muons pass through the calorimeter entering from the negative z side. This is at the origin of the slope of the time values. The visible discontinuities at $z = 0$ and $z = \pm 3000$ mm for the 2008 data are due to the uncorrected time differences between the four TileCal partitions. These were calculated using the 2008 data and adjusted for the 2009 running period. After the muon time-of-flight corrections (b), the timing shows an almost flat distribution within 2 ns in each partition, confirming a good inter-calibration between modules with the laser system. In 2009, the TOF-corrected timing distribution (c) shows an timing inter-calibration between the partitions of $\pm$ 5 ns. In preparation for the 2010 run, the 2009 single beam results were used to provide the timing calibration for all cells and, as is shown in Fig. A.3d for the 2010 single beam results, all remaining non-uniformities were corrected. The spread of the TileCal cell timing distribution at the start of the 7 TeV collisions is of order 0.5 ns.[1]

Calorimeter Performance in Collision Events

In November 2009 the LHC entered a tuning phase to reach the optimal operating conditions. The center of mass energy of the proton–proton collisions was then increased from 900 GeV, corresponding to protons colliding at the injection energy, to 2.36 TeV and finally up to 7 TeV. In Fig. A.4 the cell response spectrum is shown for collision events at different $\sqrt{s}$. The same distribution for PYTHIA MC minimum-bias simulated events at $\sqrt{s} = 7$ TeV is shown for comparison. The events were triggered using the MBTS system (Minimum Bias Trigger Scintillators) [2] both in data and MC. In order to compare the signal spectrum with the noise, the distribution resulting from random triggered events are also shown. The results show that the MC provides a good description of the response spectrum both for noise and signal regions. The response as a function of η is shown in Fig. A.5. In order to suppress the noise only the cells with a signal >500 MeV are used. The agreement is within 2 % in the central region, while is worse at large eta ($\eta > 1.2$) where also the statistics is poorer.

[1]This value takes into account 97 % of the TileCal channels.

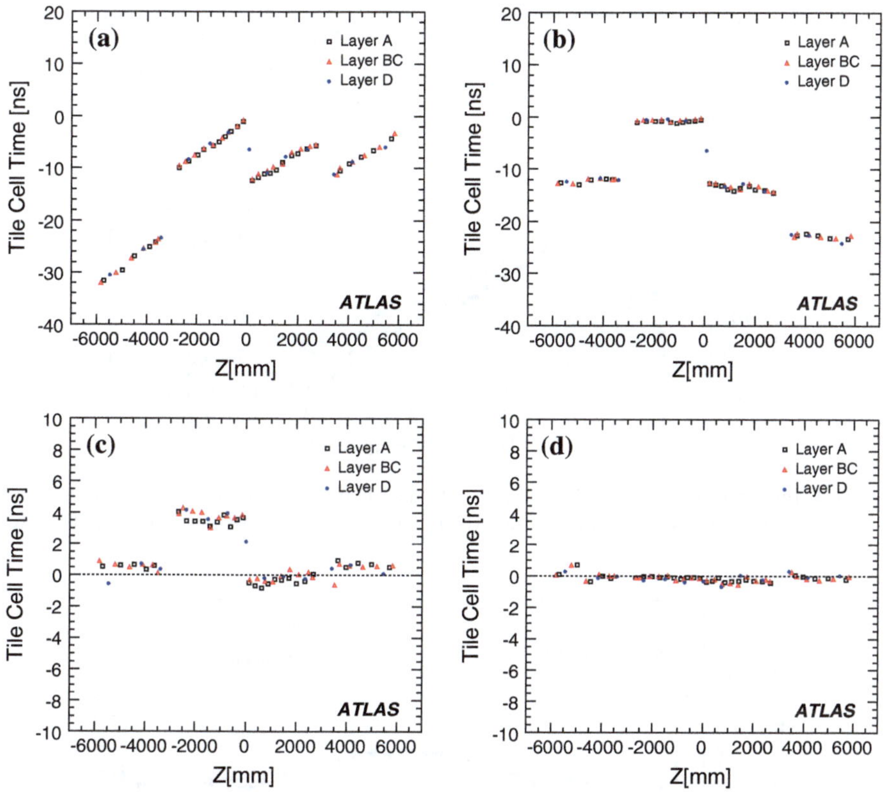

Fig. A.3 Timing of TileCal signals recorded with single beam data in September 2008 **a** and **b**, November 2009 **c** and February 2010 **d**. The time is averaged over the full range of the azimuthal angle ϕ for all cells with the same z-coordinate (along beam axis), shown separately for the three radial layers. Corrections for the muon time-of-flight along the z axis are applied in the **b**, **c** and **d** figures, but not on the *top left* **a**

Summary

All the studies presented in this appendix were essential for the commissioning of the Tile calorimeter. The results of the cosmic-muon analysis was the first study on the TileCal EM scale and the energy inter-calibration in the ATLAS cavern, and contributed to constraint the uncertainty on the energy measurement. The timing calibration with splash events was fundamental to inter-calibrate the detector with a precision of ∼0.5 ns, improving the energy signal reconstruction. Finally, the studies of the energy response in first collisions gave additional confidence on the good understanding of the calorimeter, concerning both the noise and the reconstruction of low energy deposits (few GeV).

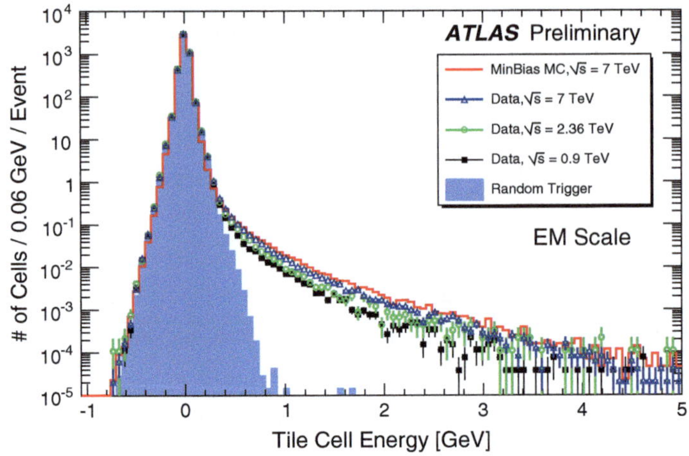

Fig. A.4 Energy response of the TileCal cells. The distributions from collision data at 7, 2.36, and 0.9 TeV are superimposed with non-diffractive minimum bias Pythia MC and randomly triggered events

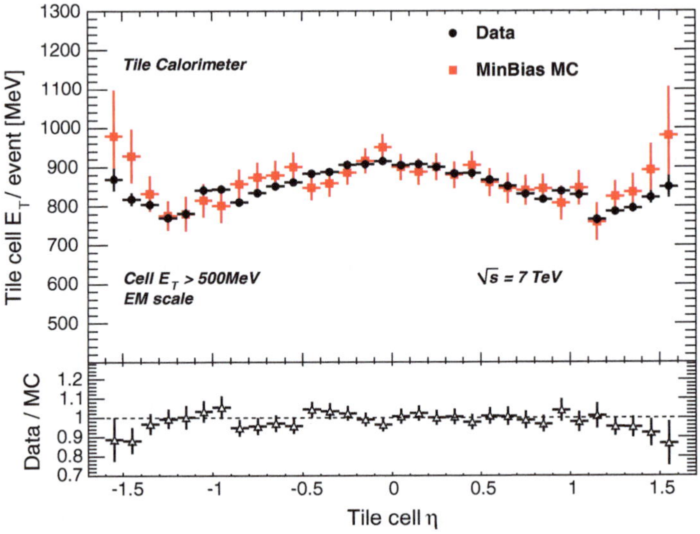

Fig. A.5 Transverse Energy (E_T) of TileCal cell as a function of eta in collision candidate events at $\sqrt{s} = 7$ TeV. Only cells with transverse energy above 500 MeV are considered. Non-diffractive minimum bias Pythia MC events with the same energy cut are superimposed with the collision candidate events. The cut is chosen at 500 MeV in order to have negligible electronic noise contribution

References

1. The ATLAS Collaboration, Readiness of the atlas tile calorimeter for lhc collisions. Eur. Phys. J. **C 70**, 1193–1236 (2010b)
2. The ATLAS Tile Calorimeter Collaboration, Tile calorimeter technical design report. CERN/LHCC 96–42 (1996)

Appendix B
Luminosity Measurement with TileCal Data

This appendix presents the main aspects of the luminosity measurement made with the Tile hadronic calorimeter (TileCal) using 7 TeV data collected in 2011. The TileCal readout provides a measurement of the energy every 25 ns. The electronic board plugged just after every PMT has also a signal integrator with an integration time (τ) ranging between 10–20 ms depending on its configuration.

Figure B.1 (left) shows the current intensities for the cells belonging to each radial layer, as a function of the pseudo-rapidity of the cell. The cells of layer A at $\eta \sim 1.3$ (called cells A13) have a higher current, since they are more exposed to particles coming from the interaction point. The integrated PMT currents are proportional to the instantaneous luminosity, and therefore can be used for luminosity measurements. The current intensity of cells A13 is plotted in Fig. B.1 (right) as a function of the luminosity measured with BCM [1], showing a good linearity between the two.

For each PMT and for each run the contribution of the noise and from beam-halo background is subtracted from the currents. This subtraction is based on the measured noise before the injection of the beams and from the integrated currents measured at the beginning of the LHC fill when the proton beams are injected, but do not yet collide. Calibration factors between each PMT current and the absolute luminosity is retrieved at the van der Meer scan (vdM scan), comparing the currents from A13 cells to the luminosity measured by BCM. Figure B.2 (left) shows the luminosity measured with TileCal cells A13 in a vdM scan in April 2011, as a function of time. The bell shapes of the luminosity is due to the movement of the beams one respect to the other in the horizontal and vertical directions. Figure B.2 (right) compares the average number of interactions per bunch crossings μ in the vdM scan measured with different sub-detectors and algorithms with to the one measured with BCM and used as default in the ATLAS collaboration. For TileCal, the luminosity is converted to μ with equation

$$L = \frac{\mu n_b f_r}{\sigma_{inel}}. \tag{B.1}$$

All the measurements are within 1 %.

V. Rossetti, *Search for Exotic Mono-jet Events*, Springer Theses,
DOI 10.1007/978-3-319-22225-7

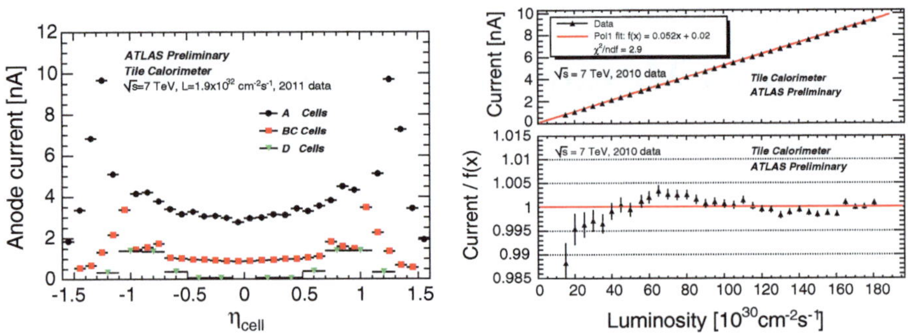

Fig. B.1 The plot on *left*-hand side shows the average current intensity for all the TileCal cells as a function of the pseudo-rapidity. The plot on *right*-hand side shows the current of cells A13 as a function of the luminosity measured with BCM

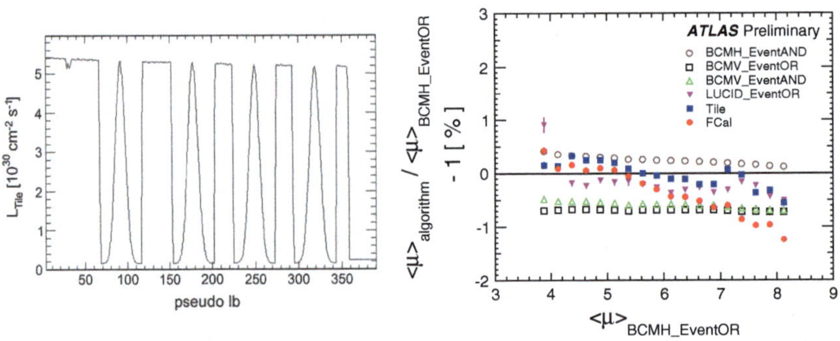

Fig. B.2 The plot on *left*-hand side shows the luminosity measured with TileCal in a vdM scan taken in April 2011. The plot on *right*-hand side compares the measured μ of different sub-detectors and algorithms for the same vdM scan

Several vdM scans have been taken in 2010, 2011 and 2012, all showing a good understanding of the TileCal measurements at low luminosity (L $\sim 10^{30}\,\mathrm{cm^{-2}\,s^{-1}}$). To study the high luminosity range, the measurements from the different sub-detectors are compared in special runs, called μ-scans. In these runs the instantaneous luminosity is varied in a large range up to L $\sim 10^{33}\,\mathrm{cm^{-2}\,s^{-1}}$. Figure B.3 presents the comparison between the luminosities measured with TileCal and the other sub-detectors, showing a good agreement between the measurements. Figure B.4 instead compares the luminosity measured in p-p runs at $\sqrt{s} = 7\,\mathrm{TeV}$ through the year 2011, showing no dependence of the measurements with time.

The luminosity studies carried out with TileCal data have been fundamental to establish the systematic uncertainty on the default luminosity measurement made by BCM. In particular the TileCal results are the only one that constrained the linearity of the luminosity measurement.

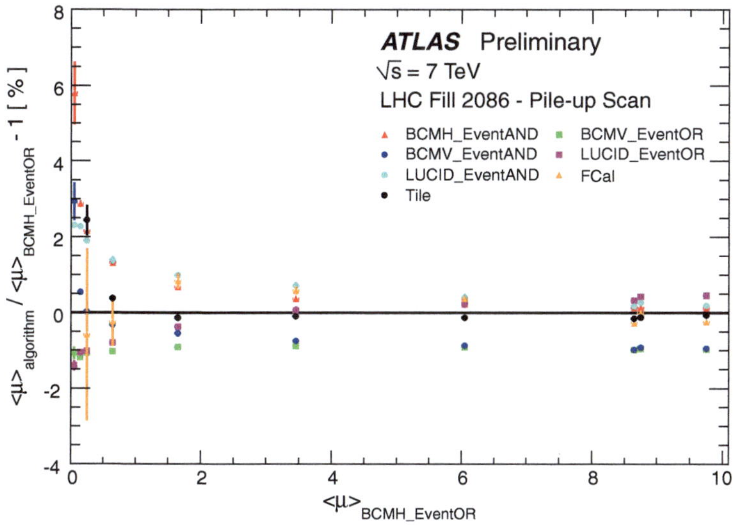

Fig. B.3 Comparison of the luminosity measurements in a μ-scan using various sub-detectors

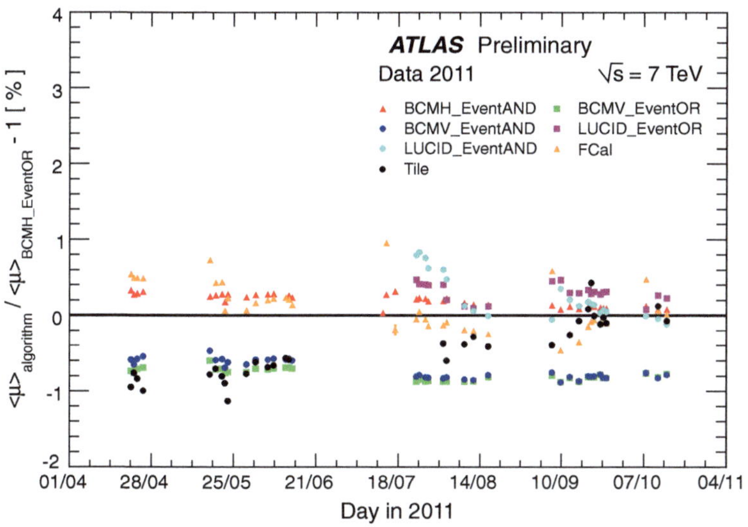

Fig. B.4 Stability of the luminosity measurements of various sub-detectors in p-p runs at $\sqrt{s} =$ 7 TeV through the year 2011

Reference

1. V. Cindro et al., The atlas beam conditions monitor. JINST **3**, P02004 (2008b)

Appendix C
Search for New Phenomena in the Mono-jet Final State with 33 pb^{-1} at 7 TeV

This appendix describes the analysis based on 33 pb^{-1} of 7 TeV collisions collected in 2010. The results of this analysis were published in Phys. Lett. B705 (2011) 294312 (see Ref. [1]), and were the first ATLAS public results on a search for new phenomena in the mono-jet channel. The analysis strategy is similar to the one adopted for the full 7 TeV dataset, therefore this appendix will only present the differences with respect to what already explained in Chap. 3.

For the analysis presented in the current appendix, the jet and electron candidates, as well as the $E_{\mathrm{T}}^{\mathrm{miss}}$, are defined with the same criteria as in the analysis of the full 7 TeV (see Sect. 3.3). Muon candidates are reconstructed with the combined algorithm and are defined by tighter kinematic requirements: $p_{\mathrm{T}} > 10$ GeV and $\eta > 2.4$.[2] Events are collected with a $E_{\mathrm{T}}^{\mathrm{miss}}$-based trigger that is 99 % efficient for $E_{\mathrm{T}}^{\mathrm{miss}} > 120$ GeV. After cleaning requirements on the vertex and on the jets, events are required to have no identified electron or muon. A first signal region (SR), called $LowPt$, is defined by requiring $E_{\mathrm{T}}^{\mathrm{miss}} > 120$ GeV, one jet with $p_{\mathrm{T}} > 120$ GeV, and no more jets with $p_{\mathrm{T}} > 30$ GeV in the event. A second SR, called $HighPt$, is instead defined by requiring $E_{\mathrm{T}}^{\mathrm{miss}} > 220$ GeV, leading jet $p_{\mathrm{T}} > 250$ GeV, second leading jet with $p_{\mathrm{T}} < 60$ GeV, and no third leading jet with $p_{\mathrm{T}} > 30$ GeV. In the $HighPt$ region, the second leading jet is required to have $\Delta\phi(jet2, E_{\mathrm{T}}^{\mathrm{miss}}) > 0.5$.

The background (BG) estimation of the Z/W+jet processes makes use of the transfer factor (TF) method described in Sect. 3.5.1. Two control regions (CRs), called inclusive electron CR and inclusive muon CR are defined by the SR selection but inverting respectively the electron and muon veto requirements. Figure C.1 shows the agreement in the number of events for the muon and electron CRs defined with different thresholds in $E_{\mathrm{T}}^{\mathrm{miss}}$ and leading jet p_{T}. The inclusive muon CR is used to estimate the $Z(\rightarrow \nu\nu)$, $W(\rightarrow \mu\nu)$, and $Z(\rightarrow \mu\mu)$ processes, while the inclusive electron CR is used for $W(\rightarrow e\nu)$, $W(\rightarrow \tau\nu)$, and $Z(\rightarrow \tau\tau)$. BG distributions for

[2] In the analysis of the full 7 TeV dataset, muons are defined with either the combined or the segment-tagged algorithms, and with $p_{\mathrm{T}} > 7$ GeV and $\eta > 2.5$.

V. Rossetti, *Search for Exotic Mono-jet Events*, Springer Theses,
DOI 10.1007/978-3-319-22225-7

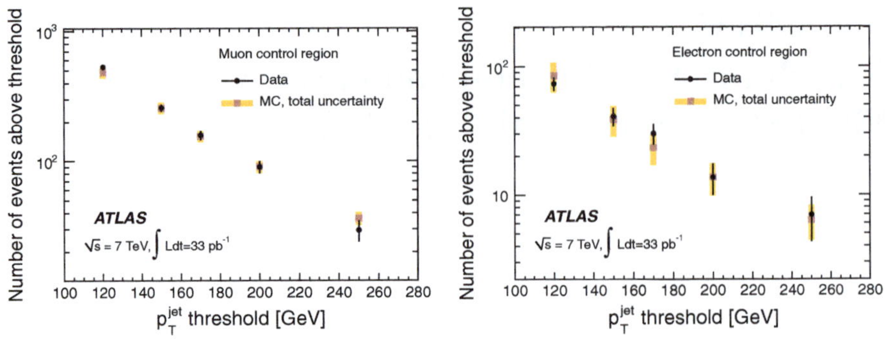

Fig. C.1 Observed number of events in the inclusive muon (*left*) and electron (*right*) CR compared to the sum of the different W/Z plus jets predictions (*squares*) as a function of the highest jet p_T threshold, in events with no second-leading jet with $p_T > 60$ GeV. The band indicates the total systematic uncertainty on the MC prediction

the SR are built from the MC-based predictions, and are normalized to the number of BG events estimated with the TF method.[3]

For the multi-jet BG estimation, we define a di-jet CR with the same selection of the SR, but requiring a second jet with $p_T > 30$ GeV and $\Delta\phi(jet2, E_T^{miss}) < 0.5$. The p_T distribution of the second leading jet is extrapolated below the 30 GeV threshold (for more details about this procedure see Sect. 3.5.2). The contamination of non-collision BGs is estimated from un-pair proton bunches in the collider that fulfill the SR selection.

Table C.1 summarizes the results with the number of observed events in the two SRs and the estimated BG, including statistical and systematic uncertainties. The statistical uncertainty is due to the limited MC statistic. The dominant systematic uncertainties is instead come from the limited statistic in the data CRs. The systematic uncertainties on $W(\rightarrow \mu\nu)$ +jets, $Z(\rightarrow \mu\mu)$+jets, and $Z(\rightarrow \nu\nu)$+jets predictions are fully correlated. Similarly, the systematic uncertainties on $W(\rightarrow e\nu)$+jets, $W(\rightarrow \tau\nu)$+jets, and $Z(\rightarrow \tau\tau)$+jets are fully correlated. The observed number of events is in good agreement with the SM predictions. Figure C.2 shows the E_T^{miss} and leading jet p_T distributions for the two SRs. Good agreement is observed in all cases. The results of χ^2 tests performed on the distributions of Fig. C.2 lead to χ^2 per degree of freedom values in the range between 0.4 and 1.2.

Results are translated into limits on the ADD model and into model independent limits on any beyond the Standard Model process. The 95 % CL upper limits on $\sigma \times A \times \epsilon$ are calculated considering the systematic uncertainties on the BG and on the integrated luminosity. The resulting values are 3.26 and 0.51 pb for the LowPt and HighPt analysis respectively.

For the graviton production in the ADD model, signal cross section and acceptances are calculated with a low energy effective field theory (see Sect. 1.4.1) with

[3]Instead in the analysis of the full 7 TeV dataset the SR distributions are built from CR data distribution corrected with bin-by-bin TFs.

Table C.1 Number of observed events and predicted BG events, including statistical and systematic uncertainties

Background predictions ± (stat.) ± (syst.)		
	LowPt selection	HighPt selection
$Z(\to \nu\nu)$+jets	$357 \pm 12 \pm 25$	$25.4 \pm 2.6 \pm 2.8$
$W(\to \tau\nu)$+jets	$139 \pm 5 \pm 36$	$7.8 \pm 1 \pm 2.3$
$W(\to \mu\nu)$ +jets	$70 \pm 4 \pm 5$	$3.8 \pm 0.6 \pm 0.4$
$W(\to e\nu)$ +jets	$59 \pm 3 \pm 15$	$3.0 \pm 0.7 \pm 0.9$
Multi-jets	$24 \pm 5 \pm 14$	–
$Z(\to \tau\tau)$+jets	$2.6 \pm 0.5 \pm 0.7$	–
$Z(\to \mu\mu)$+jets	$1.9 \pm 0.4 \pm 0.1$	–
Top	$0.96 \pm 0.04 \pm 0.2$	–
γ+jets	$0.35 \pm 0.17 \pm 0.5$	–
$Z(\to ee)$+jets	–	–
Non-collision background	$2.4 \pm 0.5 \pm 1.1$	–
Total background	$657 \pm 15 \pm 62$	$40 \pm 2.9 \pm 4.8$
Events in data (33 pb^{-1})	611	39

The statistical uncertainties are due to limited MC statistics. The dominant systematic uncertainties come from the limited statistics in the data CRs

energy scale M_D. MC signal samples are produced with PYTHIA as it is done for the analysis of the full 7 TeV dataset (see Sect. 3.7). To obtain limits on the ADD parameter M_D, model-dependent uncertainties on the signal cross sections and acceptances are determined and they are all included in the limit calculation. Systematic uncertainties affecting the production cross section include the PDF uncertainty (evaluated using the variations between the nominal CTEQ6.6 and the 44 error sets) and the uncertainty on the renormalization and factorization scales (estimated by varying them by a factor of two). Systematic uncertainties that affect the signal acceptance are evaluated on the ISR and FSR modeling (by varying the parameters controlling ISR/FSR within a range that is consistent with experimental data), jet energy scale and resolution, pile-up uncertainty (by comparing MC results with and without overlaid pile-up collisions), and luminosity. The total systematic uncertainty is 20 % for both LowPt and HighPt regions, dominated by the ISR/FSR uncertainty (13 %). An estimate of the relative importance of the signal predictions in the unknown ultra-violet kinematic region can be made by evaluating the cross section after rejecting events for which $\hat{s} > M_D^2$. For the HighPt selection, this truncation results in a change of event yield of 2, 28, and 60, respectively for n = 2, 4, and 6. For this reason only limits on n = 2, 3 and 4 were considered.

Since the HighPt selections provide the best expected limits, they are used to set the observed limits. Signal cross sections times acceptance predicted by the effective theory for n = 2 and 4 are shown on the left side of Fig. C.3 as a function of M_D. The bands around the curves reflect the uncertainties described previously. The cross

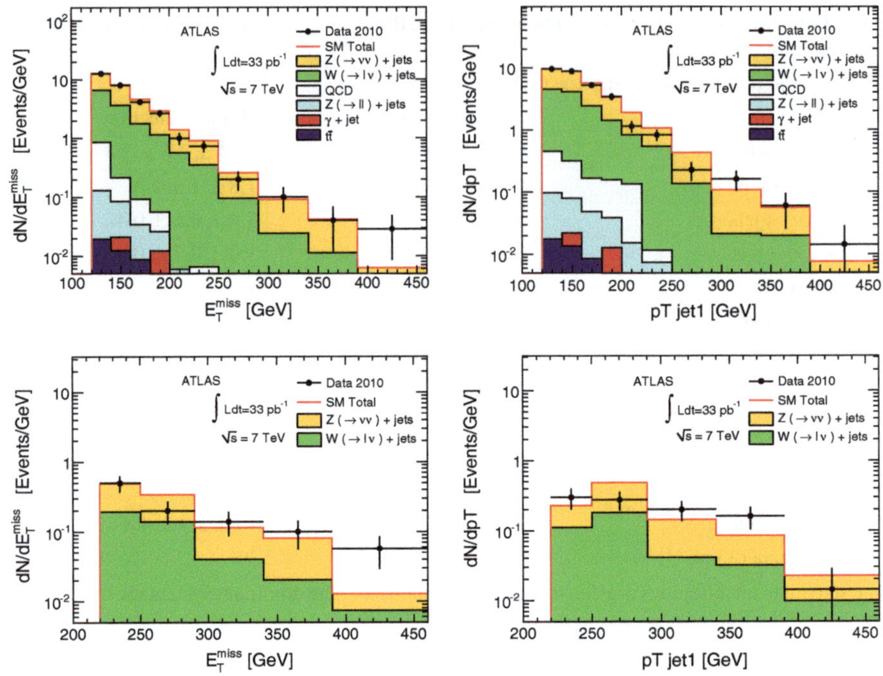

Fig. C.2 Measured $E_{\mathrm{T}}^{\mathrm{miss}}$ (*left*) and leading jet p_{T} (*right*) distributions for the LowPt (*top*) and HighPt (*bottom*) SRs compared to BG predictions. Only statistical uncertainties on the data are shown. The systematic uncertainties on the total number of predicted events are 9 % for the LowPt region and 12 % for the HighPt region

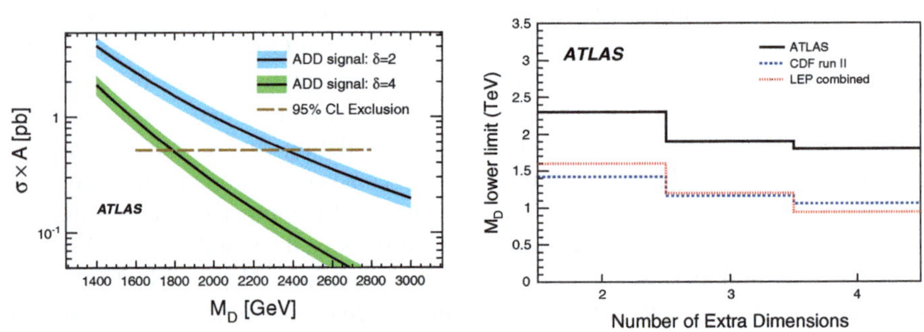

Fig. C.3 *Left* Signal cross section times acceptance as a function of M_D predicted by the effective ADD theory for 2 and 4 extra dimensions. The bands surrounding the *curves* reflect the systematic uncertainties. The observed limit is shown as a *dashed line*. *Right* 95 % CL observed lower limits on MD for different numbers of extra dimensions for ATLAS, CDF, and LEP

section times acceptance limit of 0.51 pb is also shown for illustrative purposes. The 95 % CL observed limits on M_D are shown on the right side of Fig. C.3, that largely extend already those from Tevatron and LEP.

Reference

1. G. Aad et al., Search for new phenomena with the monojet and missing transverse momentum signature using the ATLAS detector in $\sqrt{s} = 7$ TeV proton–proton collisions. Phys. Lett. B **705**, 294–312 (2011). doi:10.1016/j.physletb.2011.10.006

Appendix D
Search for New Phenomena in the Mono-jet Final State with 1 fb^{-1} at 7 TeV

This appendix presents the analysis based on the first 1 fb^{-1} of 7 TeV collisions collected in 2011. The results of this analysis were documented in the ATLAS public note ATLAS-CONF-2011-096 [1].

The analysis strategy follows closely the one used for the first 33 pb^{-1} of 2010 data and described in the previous appendix. One difference is due to an electronic failure in the LAr calorimeter affecting part of the data, and referred as "LAr hole" (see Sect. 3.1). Dedicated cleaning requirements were implemented in order to remove events affected by this failure. After this cleaning the impact of the LAr hole in the analysis is estimated to be negligible. A second difference is the additional SR, called "veryHighPt", that requires the same selection of the "HighPt" SR with increased cuts ($E_{\text{T}}^{\text{miss}} > 300\,\text{GeV}$ and leading jet $p_{\text{T}} > 350\,\text{GeV}$) to select events in tail of the $E_{\text{T}}^{\text{miss}}$ distribution.

The Z/W+jet BG estimation is based on the transfer factor method and on the inclusive muon and electron CRs. Figure D.1 shows the agreement of the CR number of events in data and MC predictions for different $E_{\text{T}}^{\text{miss}}$ and leading jet p_{T} thresholds. Also the other BG processes are estimated as described in Appendix C, with the exception of the non-collision BG. Due to an enhancement of the rate of beam-halo muons in the calorimeters, the non-collision BG is estimated using the same procedure adopted for the full 7 TeV dataset (see Sect. 3.5.4), and it constitutes at most a 2 % of the total BG.

Results are presented in Table D.1, that compares the observed number of events with the expected BGs in the three SRs. Figure D.2 shows the $E_{\text{T}}^{\text{miss}}$ and leading jet p_{T} distributions for the LowPt and veryHighPt SRs. The observation in data agrees with the estimated BG, and the results are translate into limits for the ADD model, as well as model independent limits on any beyond the Standard Model process.

The limit setting procedure is the same as for the analysis of the first 33 pb^{-1}. The 95 % CL upper limits on $\sigma \times A \times \epsilon$ for the LowPt, HighPt, and veryHighPt SRs are 1.7, 0.11, and 0.035 pb, respectively. Signal cross sections times acceptance predicted by the effective theory for n=2 and 4 are shown on the left side of Fig. D.3 as a function

V. Rossetti, *Search for Exotic Mono-jet Events*, Springer Theses,
DOI 10.1007/978-3-319-22225-7

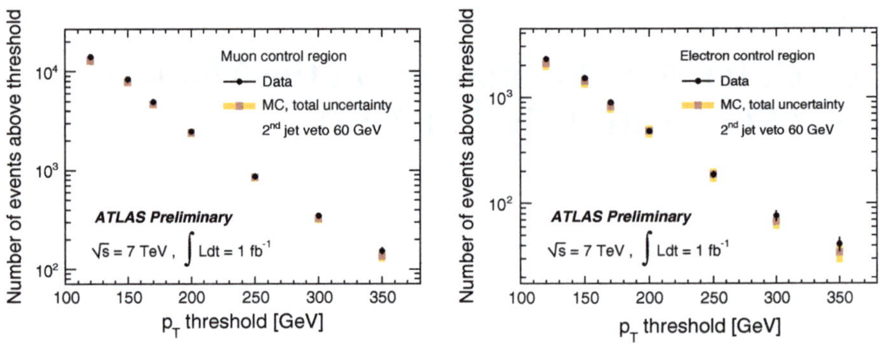

Fig. D.1 Observed number of events in the inclusive muon (*left*) and electron (*right*) CR compared to the sum of the different W/Z plus jets predictions (*squares*) as a function of the highest jet p_T threshold, in events with no second-leading jet with $p_T > 60$ GeV. The band indicates the total systematic uncertainty on the MC prediction

Table D.1 Number of observed events and predicted BG events, including statistical and systematic uncertainties

Background predictions ± (stat.) ± (syst.)			
	$LowPt$ selection	$HighPt$ selection	$veryHighPt$ selection
$Z(\rightarrow \nu\nu)$+jets	$7700 \pm 90 \pm 400$	$610 \pm 27 \pm 47$	$124 \pm 12 \pm 15$
$W(\rightarrow \tau\nu)$+jets	$3300 \pm 90 \pm 220$	$180 \pm 16 \pm 22$	$36 \pm 7 \pm 8$
$W(\rightarrow \mu\nu)$ +jets	$1890 \pm 70 \pm 100$	$113 \pm 14 \pm 9$	$18 \pm 4 \pm 2$
$W(\rightarrow e\nu)$ +jets	$1370 \pm 60 \pm 90$	$68 \pm 10 \pm 8$	$8 \pm 1 \pm 2$
Multi-jets	$360 \pm 20 \pm 290$	$30 \pm 6 \pm 11$	$3 \pm 2 \pm 2$
$Z(\rightarrow \tau\tau)$+jets	$59 \pm 3 \pm 4$	$2.0 \pm 0.6 \pm 0.2$	–
$Z(\rightarrow \mu\mu)$+jets	$45 \pm 3 \pm 2$	$2.0 \pm 0.6 \pm 0.1$	–
$t\bar{t}$	$17 \pm 1 \pm 3$	$1.7 \pm 0.3 \pm 0.3$	–
Non-collision BG	$370 \pm 40 \pm 170$	$8.0 \pm 3.3 \pm 4.1$	$4.0 \pm 3.2 \pm 2.1$
Total background	$15{,}100 \pm 170 \pm 680$	$1010 \pm 37 \pm 65$	$193 \pm 15 \pm 20$
Events in data	15,740	965	167

The statistical uncertainties are due to limited MC statistics. The dominant systematic uncertainties come from the limited statistics in the data control regions. The systematic uncertainties on $W(\rightarrow \mu\nu)$ +jets, $Z(\rightarrow \mu\mu)$+jets, and $Z(\rightarrow \nu\nu)$+jets predictions are fully correlated. Similarly, the systematic uncertainties on $W(\rightarrow e\nu)$ +jets, $W(\rightarrow \tau\nu)$+jets, and $Z(\rightarrow \tau\tau)$+jets are fully correlated

of M_D. For the ADD model the limits are extracted for n ranging between 2 and 6, and are extracted from the HighPt results to minimize the effects of the phase space for which $\hat{s} > M_D^2$. The 95 % CL observed limits on M_D are shown on the right side of Fig. C.3, that largely extend those from the previous ATLAS results.

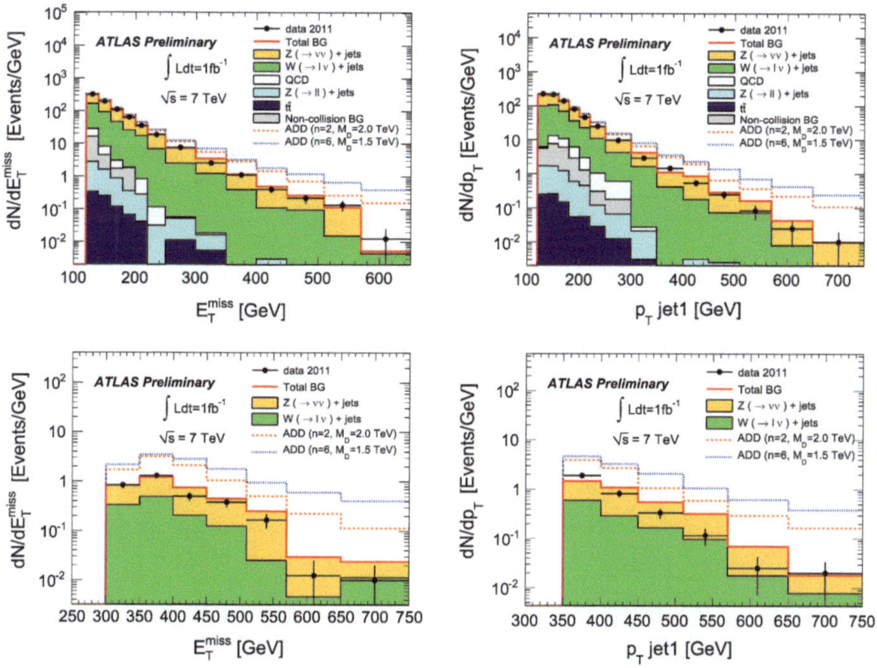

Fig. D.2 Measured $E_{\mathrm{T}}^{\mathrm{miss}}$ (*left*) and leading jet p_{T} (*right*) distributions for the LowPt (*top*) and HighPt (*bottom*) SRs compared to BG predictions. Only statistical uncertainties on the data are shown. The systematic uncertainties on the total number of predicted events are 9 % for the LowPt region and 12 % for the HighPt region

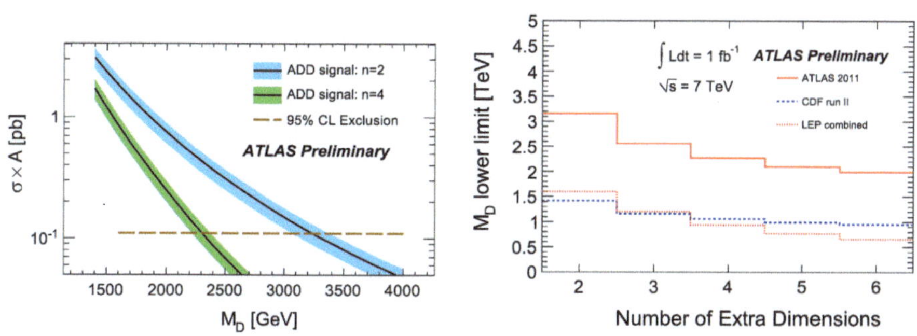

Fig. D.3 *Left* Signal cross section times acceptance as a function of M_D predicted by the effective ADD theory for 2 and 4 extra dimensions. The bands surrounding the *curves* reflect the systematic uncertainties. The observed limit is shown as a *dashed line*. *Right* 95 % CL observed lower limits on M_D for different numbers of extra dimensions, compared with previous results from ATLAS, CDF and LEP

Reference

1. Search for new phenomena in monojet plus missing transverse momentum final states using 1 fb^{-1} with the atlas detector (ATLAS-CONF-2011-096) (2011)

Appendix E
Notes on the Z/W+jets Background Estimation

This appendix describes an alternative estimation of the Z/W+jets background events for the analysis of the 7 TeV dataset. The results of this alternative estimation were taken as default for the paper [1], and were used for the limit settings in Sect. 3.7. This estimation follows closely the method presented in Sect. 3.5.1, with some differences that are explained in the following. As already mentioned, the two methods lead to almost identical results.

Z/W+jets processes are estimated from four control regions (CRs) defined to select separately $W(\rightarrow \mu\nu)$, $Z(\rightarrow \mu\mu)$, $W(\rightarrow e\nu)$ and $Z(\rightarrow ee)$ events. The correspondence between CRs and SR background processes is shown in Table E.1.

Control regions with identified muons In this alternative procedure the CRs with muons ($W(\rightarrow \mu\nu)$ and $Z(\rightarrow \mu\mu)$) are defined with tighter requirements. In particular muons are required to have $p_{\mathrm{T}} > 20$ GeV, $|\eta| < 2.4$ and $|z0| < 10$ mm, where $z0$ is the impact parameter along z with respect to the reconstructed primary vertex. Furthermore muons are also required to have the scalar pT sum of tracks within R = 0.2 around the muon track, excluding the muon itself, must be less than 10 % of the muon pT. Additionally, for the $W(\rightarrow \mu\nu)$ CR the events are required to have $E_T^{miss,\mu} > 25$ GeV, where $E^{miss,\mu}$ is the missing transverse energy as defined in Sect. 3.4, considering also the muon 4-momentum in the computation. These tighter requirements reduce the number of selected events and therefore increase the statistical uncertainty of the BG estimation. All other requirements on the reconstructed boson mass and on the muons are the same respect to the estimation described in Sect. 3.4.

Control regions with identified electrons The CRs $W(\rightarrow e\nu)$ and $Z(\rightarrow ee)$ CRs are used to estimate $Z(\rightarrow \nu\nu)$ and $W(\rightarrow e\nu)$ contributions. Events are selected using a single electron trigger and electrons are selected with $p_{\mathrm{T}} > 25$ GeV to be fully efficient. For the $W(\rightarrow e\nu)$ CR exactly one electron and $40 < M_T < 100$ GeV is required. Instead for the $Z(\rightarrow ee)$ CR the events are required to have exactly two electrons and $76 < M_{ee} < 116$ GeV. As mentioned earlier, the same selection

V. Rossetti, *Search for Exotic Mono-jet Events*, Springer Theses,
DOI 10.1007/978-3-319-22225-7

Table E.1 Overview of processes in the control regions (CR) used to estimate background contributions of processes in the signal regions (SR)

SR	$Z \to \nu\bar{\nu}$+jet	$W \to \tau\nu$+jet $W \to \mu\nu$+jet	$W \to e\nu$+jet	$Z \to ø^{+}ø^{-}$+jet $Z \to {}^{-+--}$+jet
CR	$W \to e\nu$+jet $W \to \mu\nu$+jet $Z \to e^{+}e^{-}$+jet $Z \to {}^{-+--}$+jet	$W \to \mu\nu$+jet	$W \to e\nu$+jet	$Z \to {}^{-+--}$+jet

criteria on $E_{\mathrm{T}}^{\mathrm{miss}}$ and the jets are applied in the CRs as in the SR. However, when the $W(\to e\nu)$ and $Z(\to ee)$ CRs are used to estimate the contribution of $Z(\to \nu\nu)$ to each SR, the $E_{\mathrm{T}}^{\mathrm{miss}}$ is substituted by $E_T^{miss,\not{e}}$ to mimic the kinematics of the decay of the Z boson to two undetected neutrinos. The standard calorimeter-based $E_{\mathrm{T}}^{\mathrm{miss}}$ is used for the CR to estimate the $W(\to e\nu)$ contribution to the SRs.

The $Z(\to \nu\nu)$ estimation The $Z(\to \nu\nu)$ contribution in the SR is estimated from all the 4 CRs, giving always compatible predictions. The final estimation is taken from an error-weighted average of the 4 estimations that takes into account both systematic and statistical errors with their correlations. The results are shown in Table E.2.

Systematic uncertainty The systematic uncertainties is evaluated as in the main method detailed in Sect. 3.5.1, with the only exemption of the uncertainty on jet energy scale (JES). In this alternative method this uncertainty is evaluated moving up and down the p_{T} of all the jets with $p_{\mathrm{T}} > 20$ GeV according to their JES uncertainty. This estimation gives lower uncertainty respect to the one given in Sect. 3.5.1 and is therefore less conservative.

Final results Table E.2 summaries all the estimation made with the alternative method explained in the appendix. The results are very close to those presented in Table 3.6 for the main method. It has to be noticed that the two methods differ for minor aspects, and therefore their results are highly correlated. Figure E.1 shows the distributions of the $E_{\mathrm{T}}^{\mathrm{miss}}$ and the p_{T} of the 2 leading jets for region 1 and 4. No significant differences are found in the shape of the distributions predicted by the two different methods.

Table E.2 Summary of background estimations for the four signal regions

Background predictions ± (data stat.) ± (MC stat.) ± (syst.)				
	Signal region 1	Signal region 2	Signal region 3	Signal region 4
$Z\nu\nu$ ($W\mu\nu$)	63,000 ± 400 ± 300 ± 2100	5410 ± 100 ± 70 ± 230	510 ± 30 ± 20 ± 30	59 ± 9 ± 6 ± 5
$Z\nu\nu$ ($Z\mu\mu$)	63,100 ± 900 ± 500 ± 2200	5100 ± 200 ± 100 ± 300	500 ± 80 ± 40 ± 50	75 ± 32 ± 16 ± 6
$Z\nu\nu$ ($We\nu$)	6200 ± 400 ± 500 ± 4000	5300 ± 120 ± 90 ± 310	510 ± 30 ± 30 ± 30	53 ± 10 ± 8 ± 4
$Z\nu\nu$ (Zee)	63,200 ± 900 ± 500 ± 3300	5300 ± 200 ± 100 ± 400	450 ± 60 ± 30 ± 30	72 ± 26 ± 10 ± 5
$Z\nu\nu$ (Comb all)	63,000 ± 400 ± 300 ± 2300	5320 ± 70 ± 60 ± 270	500 ± 20 ± 20 ± 30	58 ± 7 ± 5 ± 4
$W\tau\nu$	31,400 ± 200 ± 200 ± 1000	1850 ± 40 ± 30 ± 70	133 ± 8 ± 7 ± 7	13 ± 2 ± 2 ± 1
$We\nu$	14,600 ± 200 ± 100 ± 500	680 ± 30 ± 20 ± 30	40 ± 6 ± 4 ± 3	5 ± 2 ± 1 ± 1
$W\mu\nu$	11,070 ± 60 ± 80 ± 560	700 ± 10 ± 20 ± 60	55 ± 3 ± 3 ± 4	6 ± 1 ± 1 ± –
$Z\tau\tau$	421 ± 7 ± 9 ± 22	15 ± 1 ± 2 ± 1	2 ± – ± 1 ± 1	–
$Z\mu\mu$	204 ± 3 ± 10 ± 16	8 ± – ± 2 ± 3	–	–
Multi-jets	1100 ± 33 ± – ± 940	64 ± 8 ± – ± 64	8 ± 3 ± – ± 8	–
$t\bar{t}$+single t	1240 ± – ± 10 ± 250	57 ± – ± 3 ± 12	4 ± – ± 1 ± 1	–
Di-bosons	302 ± – ± 5 ± 61	29 ± – ± 1 ± 5	5 ± – ± 1 ± 1	1 ± – ± – ± –
NCB	560 ± 60 ± – ± 60	25 ± 13 ± – ± 3	–	–
Estimated background	124,000 ± 600 ± 500 ± 4400	8800 ± 100 ± 100 ± 400	750 ± 30 ± 20 ± 50	83 ± 10 ± 7 ± 6
Data	124,703	8631	785	77

The estimation of the Z/W+jets background is done with the method detailed in this Appendix E. The $Z(\rightarrow \nu\nu)$ estimation used in the background sum corresponds to the combination of the results obtained with the four control regions. In cases where both data and MCs play a role in the statistical errors, the two contributions are shown separated as second and third uncertainties. The last quoted uncertainty is the systematic uncertainty. The number of observed data events was cross checked by different analysis teams, that converged on the same numbers to the level of 1/10,000

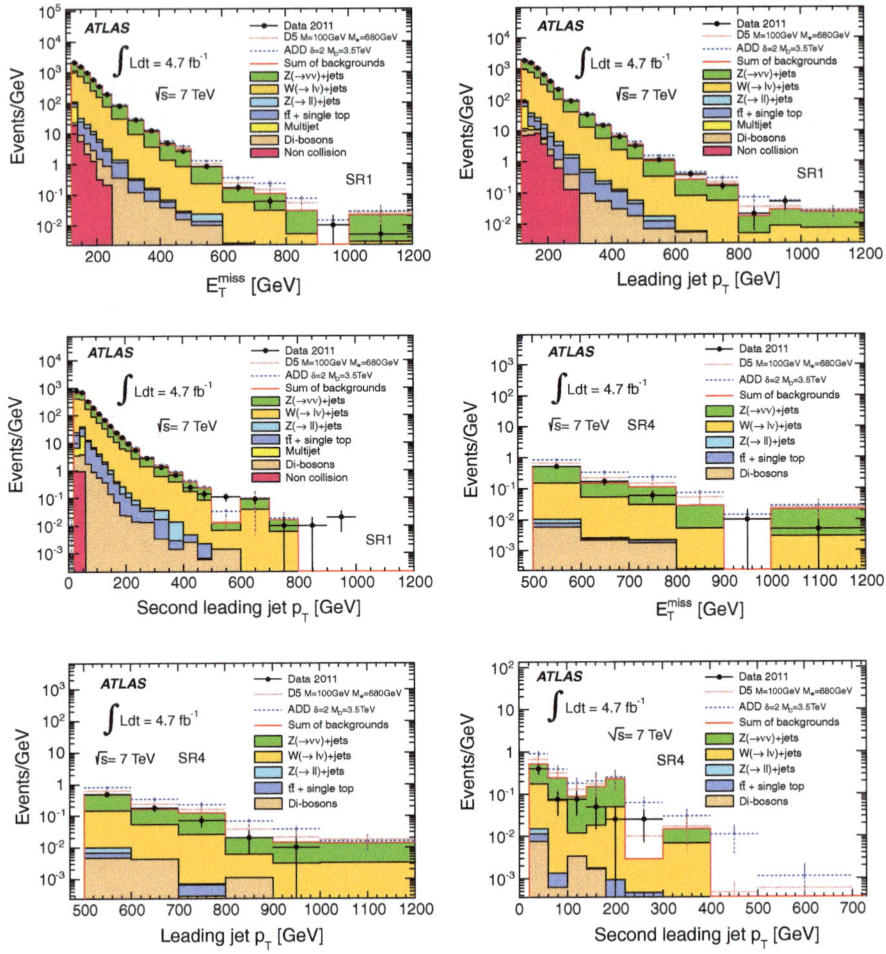

Fig. E.1 Kinematic distributions of data and estimated BG in the signal regions SR1 and SR4. This are the results of the alternative method described in this appendix

Reference

1. G. Aad et al., Search for dark matter candidates and large extra dimensions in events with a jet and missing transverse momentum with the ATLAS detector. JHEP **04** (2013)

Appendix F
Search for New Phenomena in the Mono-photon Final State with 4.6 fb^{-1} at 7 TeV

This appendix reports the main results of the search for new phenomena in the mono-photon final state (large missing transverse momentum and one high p_T photon) based on 4.6 fb^{-1} at 7 TeV. This analysis has been published in Phys. Rev. Lett. 110 (2013) 011802 [1]. The mono-photon analysis was led by the IFAE team in ATLAS, and is presented here for completeness, although it is not strictly part of this thesis. This analysis is complementary to the mono-jet search, and could be regarded as an independent cross check in case a signal for new physics would appear in the mono-jet channel.

In the mono-photon analysis, data are collected using a E_T^{miss}-based trigger. Events are required to have $E_T^{miss} > 150$ GeV, and a photon is with $p_T > 150$ GeV and $|\eta| < 2.37$. To increase the signal acceptance and reduce systematic uncertainties related to the modeling of ISR, events are allowed to have one jet (with $p_T > 30$ GeV and $|\eta| < 4.5$). The reconstructed photon, E_T^{miss} vector and jets (if found) are required to be well separated in the transverse plane with $\Delta\phi(\gamma, E_T^{miss}) > 0.4$, $\Delta R(\gamma, \text{jet}) > 0.4$, and $\Delta\phi(\text{jet}, E_T^{miss}) > 0.4$. Events with identified electrons or muons are vetoed to reduce the background from Z/W+jets and Z/W+photon.

The background estimation is very close to the one used in mono-jet analyses. One control regions is defined by the presence of a photon and a muon, and is employed for the estimation of the photon+Z/W processes. In addition, other control regions with identified leptons are used to estimate the Z/W+jets processes, with jets or electrons faking photons. The data is compared with the SM predictions in Fig. F.1 and Table F.1. A good agreement is found both in the number of events and the shape of the E_T^{miss} and photon p_T distributions.

The results are translated into exclusion limits in the context of the ADD model and WIMP pair production, following the same limit setting used in the mono-jet analyses. Figure F.2 present the 95 % CL limits on M_D, while Fig. F.3 present the 90 % CL limits on the WIMP-nucleon cross section, for spin-dependent and spin-independent interactions. As expected from the differences between strong and electromagnetic strengths, the mono-photon limits in both interpretations are less restrictive compared to the mono-jet results.

V. Rossetti, *Search for Exotic Mono-jet Events*, Springer Theses,
DOI 10.1007/978-3-319-22225-7

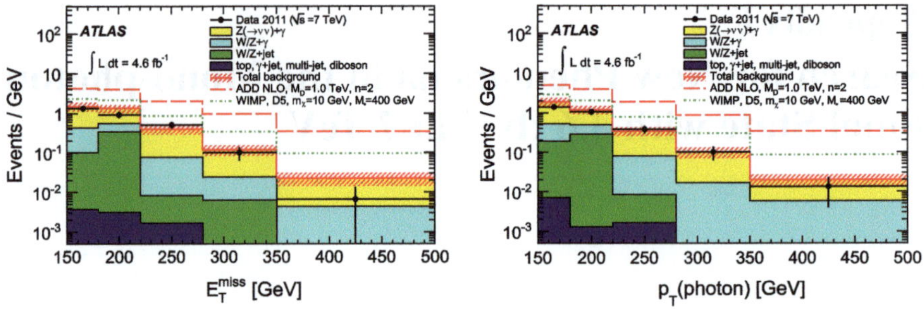

Fig. F.1 $E_{\mathrm{T}}^{\mathrm{miss}}$ (*left*) and photon p_{T} (*right*) distributions in the SR. The data (*black dots*) are compared to the SM predictions (*solid lines*). The effect of a signal in two particular ADD and WIMP scenarios is also shown for illustration purposes. For data only statistical uncertainties are included. The band around the total background prediction includes uncertainties on the data-driven background estimates and statistical uncertainties on the MC samples

Table F.1 The number of events in data compared to the SM predictions, including statistical and systematic uncertainties

Background source	Prediction	±(stat.)	±(syst.)
$Z(\rightarrow \nu\nu) + \gamma$	93	±16	±8
$Z(\rightarrow \ell\ell) + \gamma$	0.4	±0.2	±0.1
$W(\rightarrow \ell\nu) + \gamma$	24	±5	±2
W/Z+jets	18	–	±6
Top	0.07	±0.07	±0.01
$WW, WZ, ZZ, \gamma\gamma$	0.3	±0.1	±0.1
γ+jets and multi-jet	1.0	–	±0.5
Total background	137	±18	±9
Events in data (4.6 fb^{-1})	116		

The quoted statistical uncertainties include both contributions from data and the limited size of the simulated samples. In the case of W/Z+jets, γ+jets and multi-jet processes a global uncertainty is quoted

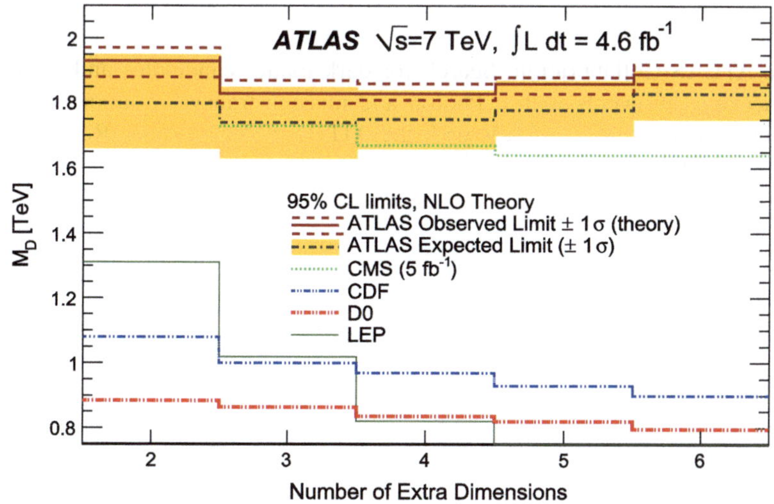

Fig. F.2 95 % CL observed (*solid lines*) and expected (*dashed-dotted lines*) limits on the scale M_D as a function of the number of extra dimensions n in the context of the ADD model. The exclusion bounds are compared with previous results (*other lines*)

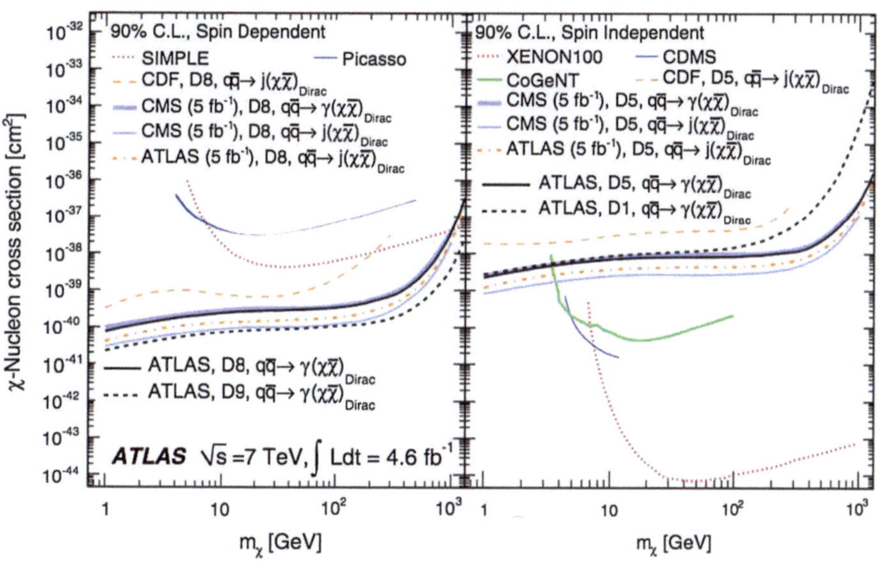

Fig. F.3 Observed exclusion limits at 90 % CL on the WIMP-nucleon cross section as a function of the WIMP mass, for spin-dependent (*left*) and spin-independent (*right*) interactions. The exclusion bounds are compared with previous results

Reference

1. The ATLAS Collaboration, Search for dark matter candidates and large extra dimensions in events with a Photon and missing transverse momentum in pp collision data at $\sqrt{s}$ = 7TeV with the ATLAS detector. Phys. Rev. Lett. **110**, 011802 (2013)

Appendix G
Search for New Phenomena in the Mono-jet Final State with 20 fb^{-1} at 8 TeV

At the time of the publication of this book, the analysis of the full 8 TeV dataset was completed, and documented in two publications [1, 2]. These include exclusion limits on the production of third generation squarks in SUSY, as well as limits on the invisibly decaying Higgs boson, in addition to the models explored previously. These analyses are not part of this thesis work. However, they are summarized in this appendix, since they constitute the most up-to-date ATLAS results on the analyses of the mono-jet final states. Moreover, the author of this thesis made important contributions also to one of these latest publications.

A first analysis was published in Phys. Rev. D. 90, 052008 (2014) [2], and was optimized for searches of stop and sbottom squarks ($\tilde{t}$ and $\tilde{b}$) production in compressed scenarios. This search focuses on the phase space in which these squarks have a similar mass ($\Delta M \lesssim 10$ GeV) with respect to the lightest neutralino χ_1^0, considered to be the lightest SUSY particle. The mono-jet final state gives a unique access to this phase space, exploiting events with one high-p_{T} jet from initial state radiation.

The analysis strategy follows closely the one used for the previous publications. The number of jets with $p_{\mathrm{T}} > 30$ GeV is constrained to be at most 3. Signal regions (SRs) are defined by progressive cuts on both $E_{\mathrm{T}}^{\mathrm{miss}}$ and leading jet p_{T}. The Z/W+jet background (BG) is estimated through orthogonal control regions (CR) with identified electrons or muons, while events with these leptons are vetoed in the SR selection.

The data is in agreement with the estimation of the SM BGs, and the results are interpreted in terms of stop pair production assuming either $\tilde{t} \to c\chi_1^0$, or $\tilde{t} \to bff'\chi_1^0$ to be the dominant decay mode. Figure G.1 shows 95 % CL exclusion limits on the stop-neutralino mass plane. In both cases, stop masses up to 260 GeV are excluded for nearly-degenerated stop and neutralino masses. The limit ranges due to the experimental and theoretical uncertainties are shown by the bands around the nominal limit lines, following the convention described in Sect. 3.7.1.

The second analysis on the mono-jet final state was submitted to EPJC for publication [1]. This analysis has few modifications compared with the previous publications: no cut is applied on the jet multiplicity, and the SRs are defined by

V. Rossetti, *Search for Exotic Mono-jet Events*, Springer Theses,
DOI 10.1007/978-3-319-22225-7

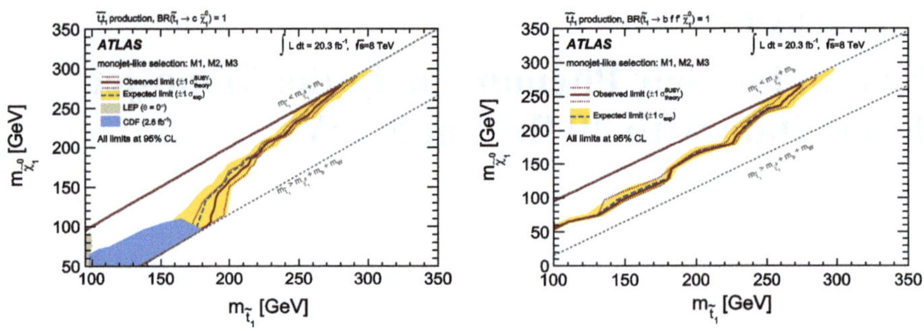

Fig. G.1 Exclusion limits at 95 % CL, in the plane of stop and neutralino masses. The limits shown on the *left* assume BR $(\tilde{t} \rightarrow c\chi_1^0) = 1$, while those on the *right* assume BR $(\tilde{t} \rightarrow bff'\chi_1^0) = 1$. The *red* and *blue-dashed lines* corresponds to observed and expected upper limits respectively, and are compared to previous results from Tevatron and LEP experiments. The *red-dotted lines* indicate the range of observed limits when varying the NLO SUSY cross sections by $\pm 1\sigma$, while the *yellow* area shows the expected $\pm 1\sigma$ ranges of limits in the absence of a signal. These results are taken from [2]

progressive thresholds on the $E_{\mathrm{T}}^{\mathrm{miss}}$ variable alone. Moreover, in order to select mono-jet topologies, the ratio between the leading jet p_{T} and the $E_{\mathrm{T}}^{\mathrm{miss}}$ is required to satisfy $p_{\mathrm{T}}/E_{\mathrm{T}}^{\mathrm{miss}} > 0.5$. Furthermore, the azimuthal separation between the $E_{\mathrm{T}}^{\mathrm{miss}}$ and any jet is required to be $\Delta\phi(jet, E_{\mathrm{T}}^{\mathrm{miss}}) > 1$, which reduces the multi-jet BG contamination. The main BG, Z($\rightarrow \nu\nu$)+jet, is estimated with a combination from four CRs, and the uncertainty on the top BG is constrained using dedicated CRs with b-tagging. These modifications result in an further reduced uncertainty on the total BG estimate: 2.7 % for $E_{\mathrm{T}}^{\mathrm{miss}} > 150$ GeV, 6.2 % for $E_{\mathrm{T}}^{\mathrm{miss}} > 500$ GeV, and 14 % for $E_{\mathrm{T}}^{\mathrm{miss}} > 700$ GeV. The data is in agreement with the SM expectations for all SRs. Figure G.2 shows the $E_{\mathrm{T}}^{\mathrm{miss}}$ and the jet-multiplicity distributions in the SR with the lowest threshold (150 GeV).

The results are translated into limits on WIMP-pair production, both in an effective-field-theory (EFT) framework (as described in Sects. 1.4.2 and 3.7.2), and considering a simplified model in which a Z' mediator couples to both quarks and WIMPs.

In the EFT approach, results are converted into limits on the WIMP-nucleon scattering cross sections, and shown in Figs. G.3a, b, as a function of the WIMP mass. The figure also includes several results from dark-matter direct-detection experiments. In addition to the operators listed in Table 3.10, the operators C1, and C5 are also considered (see [3] for the naming convention scheme). These describe the interaction between SM particles and complex-scalar WIMPs, in contrast to the D-operators which are for dirac-fermion WIMPs.

The EFT approach assumes a mediator between DM and SM particles with a mass M_{med} considerably larger then the momentum transfered in the hard interaction Q_{tr}. In this case, the mediator can not be produced directly at LHC, and can be integrated out in the EFT formalism. Using a minimal condition for the validity of

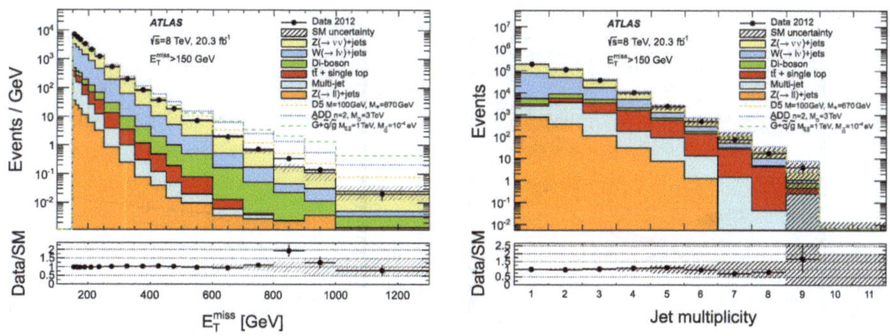

Fig. G.2 Distribution of $E_{\mathrm{T}}^{\mathrm{miss}}$ (*left*) and number of jets with $p_{\mathrm{T}} > 30$ GeV (*right*) in the SR1, which requires $E_{\mathrm{T}}^{\mathrm{miss}} > 150$ GeV . The SM background processes are estimated with a bin-by-bin data-driven procedure, and examples of signals from exotic processes are overlaid for illustration. The observed data measurement agrees with the SM expectation within the uncertainties. These results are taken from [1]

the EFT, the results in Fig. G.3a are retrieved only using the region of phase space with $Q_{tr} < M_{med}$. In order to relate M_{med} and the scale M^*, assumptions are made on the structure of the interactions for each operator, and strength of the mediator couplings with SM and DM particles, g_{SM} and g_{DM} respectively. As an example, the relation $M_{med} = M^* \sqrt{g_{SM} g_{DM}}$ is used for the D5 operator, while for the D1 operator it is considered that $M_{med} = \sqrt{y_q g_{DM}} \sqrt{M^{*3}/m_q}$, assuming a Yukawa coupling y_q for quarks of mass m_q. For each operator, two values of the couplings are chosen: 1, and the maximum value to keep the interaction perturbative. More details on the overall procedure are given in an appendix in [1].

For the operators which contribute to spin-independent interactions (Fig. G.3a) the ATLAS bounds are particularly relevant for WIMP masses below 5–10 GeV, while for spin-dependent interactions they are significantly stronger than those from direct detection experiments (Fig. G.3b) up to ~1 TeV.

To study the limitations of the EFT approach, limits are also retrieved using a simplified model with a vector mediator, which corresponds to operator D5 in Table 3.10. Figure G.3d shows the exclusion limit on M* as a function of the mediator mass and for two width values. The plot shows three regimes, depending on the mediator mass. For $M_{med} > 5$ TeV, the limits from this model coincide with those from the effective operator D5, as expected. For moderate masses (500 GeV>$M_{med} > 5$ TeV) the resonant production of the mediator results in an increase in cross section and therefore to larger constraints on M*. For $M_{med} < 500$ GeV, the production is suppressed because the mediator has a comparable or smaller mass respect to the WIMP pair. In this case, the bounds on M* using the simplified model are less stringent than those of the D5 operator.

The results of the analysis are interpreted into limits on the graviton production, in the framework of the ADD model described in Sects. 1.4.1 and 3.7.1. Figure G.4a shows the observed and expected limits on the reduced gravity-scale M_D, as a

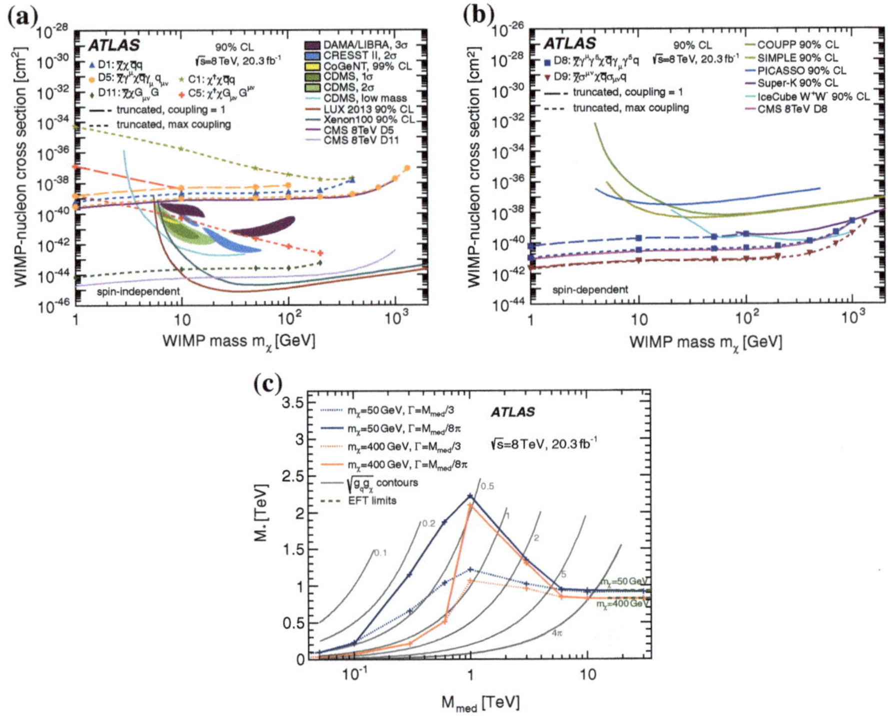

Fig. G.3 90 % CL limits on the spin-independent **a** and spin-dependent **b** WIMP-nucleon cross section as a function of WIMP mass. Results from direct-detection experiments, as well as CMS results are shown for comparison. **d** Observed 95 % CL exclusion limits on the scale M* in the case of the simplified model with a Z' mediator are shown in the *lower figure*. Two values for the WIMP masses (50 and 400 GeV) and two values of the mediator width are shown ($M_{med}/3$ and $M_{med}/8\pi$). *Grey* lines indicate where the product of the mediator couplings (to WIMPS, and to quarks) is constant. All results are taken from [1]

function of the number of extra-dimensions n. Values of M_D larger than 5.25 TeV (3.06 TeV) are excluded at 95 % CL, for $n = 2$ ($n = 6$). These bounds extend significantly those from [1], which are also shown in the plot. As discussed in Sect. 3.7.1, the signal model can not be considered fully valid for $\hat{s} > M_D^2$. In order to quantify the impact of this region of phase space on the results, limits were recomputed suppressing the cross section by a factor $M_D^4/\hat{s}^2$, for events with $\hat{s} > M_D^2$. Limits computed with this procedure are shown in Fig. G.4a by the red dashed line. For $n = 2$ this procedure has a negligible impact, while for $n = 6$ the constraint on M_D varies by 3 %.

The associated production of gravitinos and squark or gluinos has been also constrained using these mono-jet results. A description of the model is given in Sects. 1.4.3 and 4.6.1. Figure G.4b shows the lower bounds on the gravitino mass, in the case of degenerated gluino and squark masses. Gravitino masses below 3.5,

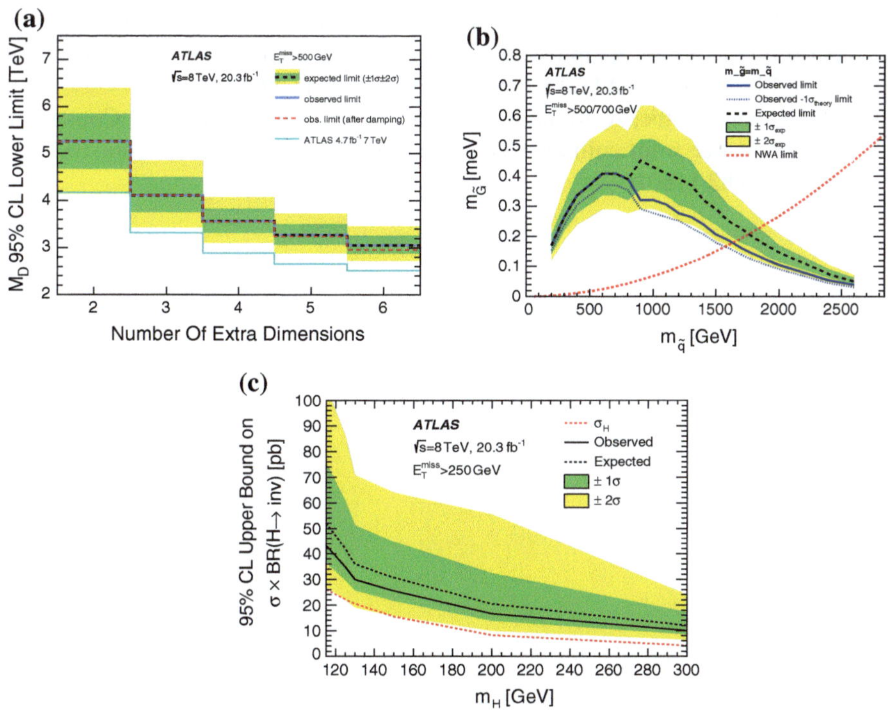

Fig. G.4 **a** Exclusion limits at 95 % CL on the scale M_D in the ADD model, as a function of the number of extra dimensions n. The *green-yellow* bands indicate the expected $\pm 1\sigma$ and $\pm 2\sigma$ ranges of limits. The same convention is used in the other two figures. **b** 95 % CL bounds on the gravitino mass as a function of the squark/gluino mass, in the degenerated case. **c** 95 % CL on the $\sigma \times$ BR (H→ inv.) as a function of the Higgs boson mass, compared with the expected production cross section of the Higgs boson assuming BR (H→inv.) = 1. All results are taken from [1]

and 2×10^4 eV, are excluded at 95 % CL for squark/gluino masses of 500 GeV, and 1.5 TeV, respectively. Instead, in the non-degenerated case exclusion limits vary between 10^4 eV, 5×10^4 eV, depending on the squark and gluino masses.

Finally, the results are interpreted as limits on the production cross section times the branching-ratio of the Higgs boson decaying to invisible final states. Figure G.4c shows the upper limits on the $\sigma \times$ BR(H→inv.) as a function of the Higgs mass. Values above 44 and 10 pb are excluded at 95 % CL, for Higgs masses of 115 and 300 GeV, respectively. The upper limits from this analysis are larger than the expectations of a Higgs boson with BR(H→inv.) = 1, which is also shown in the figure.

References

1. G. Aad et al., Search for new phenomena in final states with an energetic jet and large missing transverse momentum in pp collisions at $\sqrt{s} = 8$ TeV with the ATLAS detector. Submitted to EPJC (2015)

2. G. Aad et al., Search for pair-produced third-generation squarks decaying via charm quarks or in compressed supersymmetric scenarios in pp collisions at $\sqrt{s} = 8\,\text{TeV}$ with the ATLAS detector. Phys. Rev. D. **90** (2014)
3. J. Goodman, M. Ibe, A. Rajaraman, W. Shepherd, T.M.P. Tait et al., Constraints on dark matter from colliders. Phys. Rev. D **82**, 116010 (2010). doi:10.1103/PhysRevD.82.116010